Urban Greening Techniques
An Introduction

World Scientific Series on the Built Environment

ISSN: 2737-4831

Series Editor: Willie Chee Keong Tan
(National University of Singapore, Singapore)

Published

World Scientific Series on the Built Environment

Volume 6

Urban Greening Techniques
An Introduction

Tan Chun Liang

National University of Singapore, Singapore

World Scientific

NEW JERSEY · LONDON · SINGAPORE · BEIJING · SHANGHAI · HONG KONG · TAIPEI · CHENNAI · TOKYO

Published by

World Scientific Publishing Co. Pte. Ltd.
5 Toh Tuck Link, Singapore 596224
USA office: 27 Warren Street, Suite 401-402, Hackensack, NJ 07601
UK office: 57 Shelton Street, Covent Garden, London WC2H 9HE

Library of Congress Cataloging-in-Publication Data
Names: Tan, Chun Liang Terrence, author.
Title: Urban greening techniques : an introduction / Tan Chun Liang,
    National University of Singapore, Singapore.
Description: New Jersey : World Scientific, [2024] | Series: World Scientific series on the
    built environment, 2737-4831 ; vol. 6 | Includes bibliographical references and index.
Identifiers: LCCN 2023026013 | ISBN 9789811278372 (hardcover) |
    ISBN 9789811279010 (paperback) | ISBN 9789811278389 (ebook for institutions) |
    ISBN 9789811278396 (ebook for individuals)
Subjects: LCSH: Urban forestry. | Urban landscape architecture. | Sustainable development.
Classification: LCC SB436 .T36 2024 | DDC 635.9/77--dc23/eng/20230627
LC record available at https://lccn.loc.gov/2023026013

British Library Cataloguing-in-Publication Data
A catalogue record for this book is available from the British Library.

For any available supplementary material, please visit
https://www.worldscientific.com/worldscibooks/10.1142/13472#t=suppl

Desk Editors: Gregory Lee/Amanda Yun

Typeset by Stallion Press
Email: enquiries@stallionpress.com

This book is dedicated to my dear Aeris.

Preface

In my training as a student of architecture, I have always pursued truth and beauty to the best of my ability. After I embarked on greenery research, it became apparent to me that the wide gap between science and practice in urban greening is a major impediment towards our sustainability agenda. I believe this book is a step towards bridging that gap.

I hope readers can use this book to minimise greenwashing and to fully realise the potential of greenery in the urban landscape.

I would like to thank Professor Wong Nyuk Hien for giving me a chance to participate in research, Professor Willie Tan for providing the opportunity to write this book and Professor Tan Puay Yok for introducing me to the fascinating subject of urban greenery. I appreciate the help from students who chipped in for the research and illustrations leading to the publication of this book.

Tan Chun Liang

About the Author

Dr Tan Chun Liang is a senior lecturer in the Department of Architecture, College of Design and Engineering, National University of Singapore. He lectures for both Bachelor as well as Master of Landscape Architecture courses, including modules such as "Introductory GIS for Landscape Architecture" and "Urban Greening: Technologies and Techniques". Besides teaching, Dr Tan also served as the PI and thrust leader of the Urban Greenery Research Thrust under the NUS–JTC Industrial Infrastructure Innovation Centre. His research focus is on urban greenery systems and their impact on heat transmission, wellness and maintainability. His work has been published in numerous journals, including *Building and Environment, Landscape and Urban Planning, Energy and Buildings, Sustainable Cities and Society, Digital Landscape Architecture, Urban Climate, Solar Energy, Chinese Landscape Architecture* (中国园林), *Science of the Total Environment* and *Nature Reviews Earth & Environment.*

Contents

CHAPTER 1

Introduction

There has been an increasing sense of urgency to address the adverse impacts of urbanisation on the environment. Anthropogenic changes to the landscape, resulting in higher temperature and more frequent flood events, have real and tangible impacts on urban inhabitants. Indeed, problems arising from urban expansion have been extensively studied and well documented. The solutions to these problems, however, are less apparent.

Nevertheless, the trend towards sustainable development persists, mobilising a significant part of society to re-evaluate our choices in resource expenditure and environmental protection. This is evident across most industries and none more so than in the built sector, where vast amounts of resources are invested into the construction, operation and maintenance of buildings and infrastructure.

Many efforts at sustainable building and development have elements of greenery incorporated within them. Some examples include urban masterplans, building policies, green building rating tools and design proposals. More often than not, the reason for having greenery is to tap on some form of benefit, such as temperature reduction and water and air purification. These benefits may have been validated through scientific studies but normally stop short of recommending how practitioners can make use of these results in their design and decision-making processes. There is, in fact, an overwhelming amount of material on the benefits of greenery but a paucity of information on proper application to activate these benefits. As a result, when it comes to practical interventions, there are not many options other than to add as much greenery as possible.

This book proposes a different approach: that of ensuring not just quantity but also quality. Potential benefits and costs of different forms of greenery should be clear before making any decision. Here we tell the

1

story of urban greenery, in its many forms, associated benefits and requirements, so that we may truly maximise the potential of greenery to better the built environment for everyone.

How to use this book

This book aims to provide readers with a clear and holistic introduction to urban greenery and its relevance at the different scales of implementation, including its benefits, limitations, as well as considerations to ensure successful deployment.

Greenery is categorised into (i) City-scale, (ii) Precinct-scale and (iii) Plant scale because the benefits, requirements and instruments involved are unique to each scale (Fig. 1). Where appropriate, frameworks are included to provide readers with a clear overview of each section. Exercises are provided to reinforce learning. References to technical manuals as well as online resources have been appended. For more information, please refer to the Instructions for Supplementary Material.

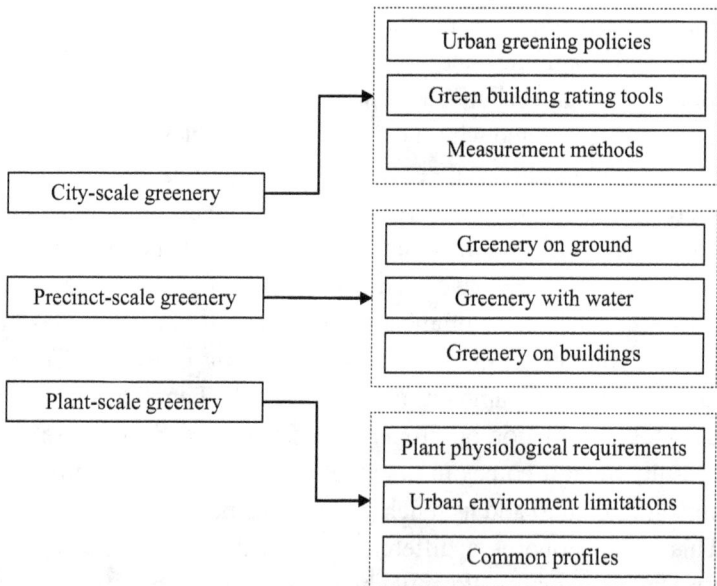

```
                                    ┌─────────────────────────────┐
                                    │   Urban greening policies   │
                                    ├─────────────────────────────┤
                              ┌────▶ │  Green building rating tools │
                              │     ├─────────────────────────────┤
                              │     │   Measurement methods       │
   ┌──────────────────┐       │     └─────────────────────────────┘
   │ City-scale greenery ├─────┘
   └──────────────────┘             ┌─────────────────────────────┐
                                    │    Greenery on ground       │
                                    ├─────────────────────────────┤
   ┌────────────────────┐   ┌─────▶ │    Greenery with water      │
   │ Precinct-scale greenery ├──┘    ├─────────────────────────────┤
   └────────────────────┘           │    Greenery on buildings    │
                                    └─────────────────────────────┘
   ┌──────────────────┐
   │ Plant-scale greenery ├──┐       ┌──────────────────────────────────┐
   └──────────────────┘    │       │ Plant physiological requirements │
                           └─────▶  ├──────────────────────────────────┤
                                    │  Urban environment limitations   │
                                    ├──────────────────────────────────┤
                                    │       Common profiles            │
                                    └──────────────────────────────────┘
```

Fig. 1.   Structure of this book.

CHAPTER 2

Urban greenery

Where does the story of urban greenery begin?

Adorning built-up areas with vegetation is certainly an age-old notion (think Hanging Gardens of Babylon), but one may argue that urban greenery has its bona fide urban planning and architectural roots in the early 20th century.

The Garden City concept, penned by Sir Ebenezer Howard (1898), describes a utopian city where people live in harmony with nature. It was a reaction to the predominantly industrial cities, plagued by issues such as overcrowding, pollution and a generally low quality of living conditions. While most of these problems remain unsolved till today, the Garden City movement has spurned countless other efforts to improve city living by introducing elements of nature into built environments (Sharifi, 2016). Greenery remains closely associated with these schemes.

Our penchant to expand settlements and then divide them into districts is only matched by an equally peculiar desire to clear natural green spaces only to reintroduce them in some form. Evidently, greenery cannot be replaced as fast and as much as it is been cleared. Urban settlements and populations continue to be on the rise: a projected 70% of the global population will live in urbanised areas, and cities will continue to expand (Kundu & Pandey, 2020). How does urban greenery factor into this statistic? An observable trend is that urban greenery is perceived to be a way to replace the loss of green cover. It is seen to be able to alleviate some of the ills of urbanised environments and provide multiple benefits to city inhabitants. Urban greenery has also evolved significantly over the years, from ground-level parks and gardens to skyrise greenery such as green wall and roof systems to address the challenges of greening highly built-up areas.

Definition and scale

The term "urban greenery" itself is rather nebulous and requires some clarification to facilitate discussion. In fact, the exact constituents of what we expect of "urban greenery" is dependent on the scale of our assessment.

At the larger scale (city, state or country), urban areas may be classified as urban, peri-urban or non-urban (rural or green) areas by observing the extent of green cover or via broader categorisations, such as population density or number of dwellings per hectare (McIntyre, Knowles-Yánez, & Hope, 2008). Such information is useful as urban planning metrics.

At the peri-urban or urban scale (e.g., a housing precinct or township), urban greenery may refer to more specific aspects of vegetation, such as type of vegetation cover, composition and configuration of green cover. It can be any object in the urban context that has vegetation. This includes forests, parks, gardens, streetscapes, waterways, trees, shrubs and turf. Clear delineation of urban and green spaces may not always be possible, as there can be overlaps (e.g., a tree in a small planter but with a large canopy over a paved area).

At the scale of a building or a garden, system and species information may be pertinent, like the species of plants used for a green wall system. Abiotic data may become relevant, such as microclimatic conditions on site or amount of daylight provision, etc.

Technically, vegetation under all three scales can be classified as urban greenery but may not be entirely relevant to the planning or design process across all scales. For instance, having landscape metric data will be of little use to the designer of a green wall. It is, therefore, useful to first identify the appropriate spatial resolution before embarking on an urban greening exercise.

More importantly, this classification can help us identify the type and extent of ecosystem services to be expected, as well as associated challenges in the design and implementation stages.

This book attempts to describe urban greenery in multiple scales, including the rural, urban periphery and fringe (peri-urban), as well as the suburban, inner urban and city centre areas (urban) (Fig. 2). The

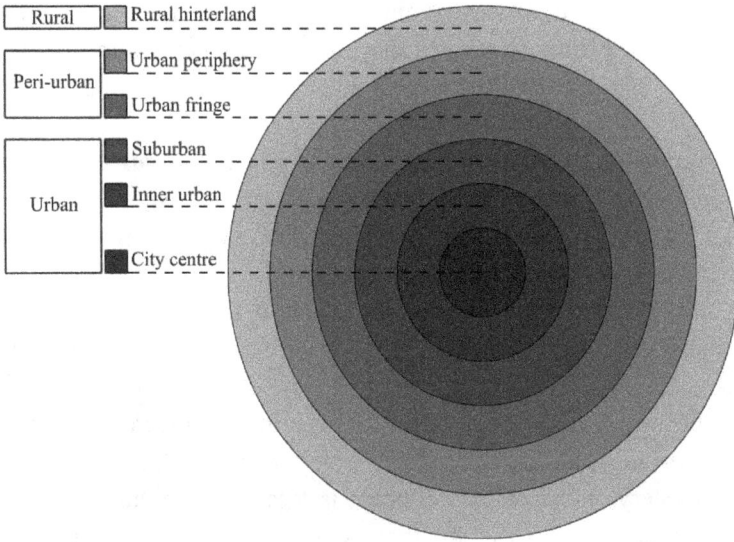

Fig. 2.   Concept of urban, peri-urban and rural areas (Piorr, Ravetz, & Tosics, 2011).

appropriate methods for assessment, design and implementation are provided accordingly.

Ecosystem services

The benefits of having greenery in the urban environment have been widely reported (Virtudes, 2016). The role of green infrastructure (employing nature-based solutions as urban infrastructure, such as sewage and water treatment) has gathered prominence over the years (Dawson et al., 2014). However, it is insufficient for practitioners to simply know about the various benefits that greenery can provide — understanding how they can be applied into design projects and their expected benefits and limitations is equally important. It is in this light that well-defined frameworks are necessary to help identify the specific services that greenery can provide and how best to realise these services.

An ecosystem can be defined as an area where living organisms interact with the environment to sustain life (Moll & Petit, 1994). Both biotic and abiotic factors can influence each other, either directly or indirectly.

For instance, a reduction in sunlight exposure will result in less photosynthetic activity, leading to lesser foliage density. This may, in turn, lessen the amount of food and shelter available for animals dependent on these plants.

The urban environment can be seen as an amalgamation of multiple individual ecosystems, coming together to form one large complex ecosystem. These ecosystems are in constant flux, ever-changing, as a result of interactions within each ecosystem or between other ecosystems.

It is within these ecosystems that we begin to find benefits that may help improve urban conditions. These benefits are known as ecosystem services. Strictly speaking, this is not a novel concept: humans have always been tapping on ecosystem services in one form or another: fish from rivers and fruits from trees are harvested for food. In the context of urban greenery, however, this concept is less intuitive, and it would be useful to adopt a design process that includes consideration of the associated ecosystem services. To do so, the concept of ecosystem services must be elaborated.

Ecosystem services can be classified into four major categories, namely, (i) Provisioning, (ii) Regulating, (iii) Supporting and (iv) Cultural services (Berghöfer et al., 2011). Some common services associated with the categories are shown in Fig. 3.

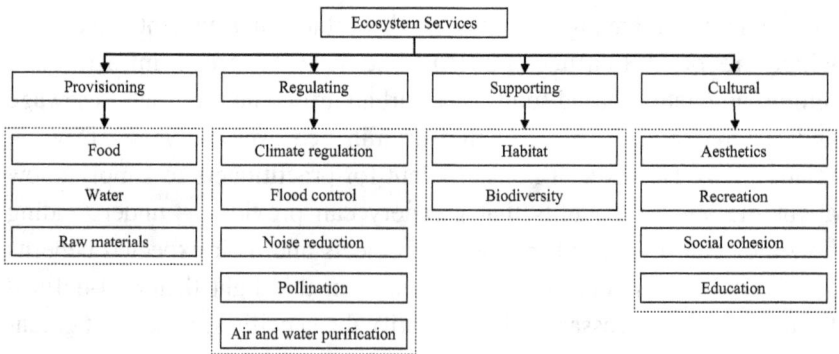

Fig. 3. Common ecosystem services.

Provisioning

Provisioning services include all items that can be obtained from ecosystems, such as food, water, medicine and raw materials. In rural and peri-urban areas, food production occurs in farms, forests, rivers, etc., within large expanses of land. Urban food production, on the other hand, tends to operate on a smaller scale and can be found in parks, gardens, backyards or even rooftops. Typically, urban areas only produce a small portion of the total demand and rely on supply from rural areas or imports (Ernstson et al., 2010). The recent COVID-19 pandemic has highlighted the urgency for cities to reconsider the issue of food security and strategies, such as introducing novel farming technologies, alternative proteins and urban farming (Khan et al., 2020; Kim, Kim, & Park, 2020). Singapore, for instance, launched a "30 by 30" initiative with the aim of producing 30% of the country's nutritional needs by 2030 through innovations in the areas of urban farming, processing technologies and alternative food sources (Mok, Tan, & Chen, 2020).

Regulating

Regulating services consist of benefits via regulation from ecosystem processes, such as climate regulation, flood control, noise reduction, pollination, and air and water purification.

Most cities suffer from increased temperatures due to the combined effects of climate change as well as the Urban Heat Island (UHI) effect (Santamouris, 2015). Urban greenery has been shown to be effective at regulating temperature fluctuations in the built environment (Wong, Tan, Kolokotsa, & Takebayashi, 2021). It is important to note that different forms of greenery can influence climate regulation in different ways. For instance, outdoor climate regulation is mainly achieved via shade provision from trees, through the reduction of direct shortwave radiation exposure. Indoor cooling, on the other hand, is more affected by the greenery system setup (either green wall or green roof) and provides thermal regulation by reducing thermal transmission into buildings and the resultant long-wave emission to occupants. The mechanisms for cooling are elaborated on in Fig. 16.

Built-up areas tend to have a higher portion of impermeable surfaces areas, leading to increase surface water runoff. This increases the risk of flooding during prolonged or extreme precipitation events. Greenery can reduce surface runoff by intercepting water through its foliage and soil. Even skyrise greenery can help to alleviate flooding by serving as water retention and detention sites during storm events. Urban landscapes with high coverage of impervious surfaces can lose about 40% to 80% of rainfall to surface runoff (Bonan, 2015).

Noise can be generated from anthropogenic activities, such as traffic and construction. These can be detrimental to human well-being, and greenery has been shown to reduce noise pollution via the reflection, refraction and absorption of sound waves (Lacasta, Peñaranda, & Cantalapiedra, 2018; Stansfeld, Haines, & Brown, 2000). Depending on the species selected and the arrangement of plants, different levels of sound attenuation can be expected (Fang & Ling, 2003).

Greenery provides food and shelter for animals such as birds and insects. When properly managed, it can help regulate pests, pollination and seed dispersal activity, maintaining healthy urban ecosystems (Buchmann & Nabhan, 2012; Ellis et al., 2005; Şekercioğlu, Daily, & Ehrlich, 2004). Different forms of greenery, such as parks, community gardens, cemeteries and roof gardens scattered throughout the city, form the bedrock for promoting pollination and mitigating habitat loss for animals due to urban expansion (Andersson, Barthel, & Ahrné, 2007; Matteson & Langellotto, 2017).

Urban areas are sources of anthropogenic activity and hotbeds of pollution, such as particulate matter arising out of carbon emissions from cars and factories and greenhouse gases like carbon dioxide, methane, chlorofluorocarbons and ozone. This leads to a degradation in environmental and human well-being (Cohen et al., 2004). Greenery can help reduce pollution levels to some extent through its biological processes, such as by capturing and storing carbon dioxide from the atmosphere in its biomass, removing harmful gases and intercepting airborne pollutions via its foliage (Escobedo, Kroeger, & Wagner, 2011; McPherson & Simpson, 1999). Contaminants in the soil and water can be extracted and removed through phytoremediation. Plants can help clean up contaminated areas and reduce the chances of contaminants spreading out to other areas (Nissim & Labrecque, 2021).

Supporting

As urban areas continue to expand and replace forested areas, an increasing number of habitats are lost. Animals are displaced by the change in land use, and this drastically affects the level of biodiversity in these areas (Werner, 2011). In this regard, greenery can help preserve biodiversity in the urban environment by providing food and shelter for animals such as insects, amphibians and birds. This is possible through increased photosynthetic activity and soil formation in green areas. Although large areas of green spaces are preferred for preserving biodiversity, skyrise greenery has also been shown to be able to support life (Oh, Richards, & Yee, 2018).

Cultural

One key benefit of having greenery in the urban environment is that it can improve the aesthetics of its surroundings, which leads to other benefits such as an increase in physical and mental health, stress reduction and other psychological benefits (Deng et al., 2020). This can have a positive impact on land value, as shown in some studies linking the price of real estate with proximity to different types of urban green spaces (Panduro & Veie, 2013; Trojanek, Gluszak, & Tanas, 2018).

In some cities, greenery has become synonymous with better standards of living for urban dwellers, and this has, in turn, resulted in more urban greening efforts. Singapore, for instance, launched the "Garden City" vision in 1967 to adorn the city-state with lush greenery in order to improve urban conditions and to make known to the world that it is a pleasant, well-organised city (Lee, 2012). Since then, Singapore's greening policy has evolved into the "City in a Garden" vision in 1998 and more recently to "City in Nature", where greening goes beyond aesthetics and focuses on aspects such as community engagement, habitat restoration and quality of life for residents (NEA, 2021).

Besides beautifying the environment, urban green spaces (UGSs) are ideal for hosting recreational activities (Konijnendijk, Annerstedt, Nielsen, & Maruthaveeran, 2013). Waterways, parks and gardens provide spaces for activities such as physical exercises, water sports, fishing, hiking, etc. The contemplative nature of parks can also help to reduce stress and improve the mental well-being of urban dwellers. As more interaction between people occurs within green spaces such as neighbourhood parks and gardens, improvements to community bonding and quality of living can be expected. This can lead to better impressions of liveability and improved social cohesion in the community. Samsudin, Yok, & Chua (2022) argued that the social capital fostered between residential communities living in dense urban settings is affected by the perceived attributes of green spaces and how they were spatially distributed in the neighbourhood.

Our understanding of the environment and how we can affect it can lead to significant changes to our attitude and behaviour towards pressing issues such as climate change and pollution. Quite often, our appreciation of these issues does not begin in the classroom but through exposure to nature and green elements in our daily lives (Wolsink, 2016). Indeed, environmental stewardship and awareness of ecosystem services are a function of exposure to greenery, such as visits to parks. Community gardens and urban farms are also great starting points for introducing concepts of biodiversity, environment and sustainability to students, as well as inculcating ecological literacy from a young age (Datta, 2016).

Ecosystem disservices

The other important (and equally compelling) reason to adopt the frameworks recommended in this book is to minimise any possible disservices that may occur due to the inclusion of greenery.

Ecosystem disservices are functions of ecosystems that are deemed to be negative for well-being and can occur in a variety of ways (Lyytimäki, Sipilä, & Greening, 2009). A park, for instance, may provide many benefits, such as thermal regulation, recreation and biodiversity preservation. But it may also become a hotbed of criminal activity under extraneous circumstances (Groff & McCord, 2012). Greenery has been widely touted to reduce urban air pollution, but they are also emitters of volatile organic compounds (VOCs) into the atmosphere (Churkina, Grote, Butler, & Lawrence, 2015). A list of disservices can be found in Table 1.

Although the benefits of greenery have been extensively highlighted in this book, it is prudent to adopt a critical and holistic approach when deciding on matters pertaining to greenery introduction.

Expectations regarding UGSs may be high due to increased publicity and awareness over the years, but it is reasonable to expect that greenery is not all good, and for every positive attribute that has been associated, there is some possible drawback that may potentially negate the benefits it can bring.

Table 1.   Examples of potential ecosystem disservices.

Type of greenery	Ecosystem service	Potential disservice	Reference
Parks	Provides spaces for recreation Promotes human well-being	Facilitates criminal activity	(Jorgensen and Anthopoulou, 2007)
Foliage	Improves air quality by trapping particulate matter	Emit biogenic volatile organic compounds (BVOCs)	(Churkina, Grote, Butler, & Lawrence, 2015)
Green spaces	Preserves biodiversity in the urban environment	Human casualties and fatalities from animal attacks	(Bombieri et al., 2018)
Trees	Shade provision Beautifies surroundings	Root systems causes damage to urban infrastructure Block views	(Lyytimäki, Petersen, Normander, and Bezák, 2008)
Flowering plants	Promotes pollination and biodiversity	Increases cases of allergic reactions in humans	(D'amato, 2000)
Ground level greenery	Reduces air temperature	Blocks wind and reduces urban ventilation	(Guo, Gao, Buccolieri, Zhang, and Shen, 2021)

Greenwashing

Global awareness of environment issues such as climate change and pollution has increased significantly in recent years, and this has given rise to calls to adopt more environmentally friendly practices for all industries. One example is for business entities to undertake Corporate Social Responsibility (CSR) practices, which entails the integration of social and environmental concerns in their business operations as well as interaction with various stakeholders (Moir, 2001). The building and construction industry has also embarked on a journey to adopt more sustainable building practices and reduce their impact on the environment. As a major source of energy consumption and pollution, initiatives to adopt green building practices and features have been met with favourable responses (Sev, 2009).

However, products and services may be marketed as environmentally friendly when, in reality, they are not. This is the basis of greenwashing, which can be defined as "the intersection of two firm behaviours: poor environmental performance and positive communication about environmental performance" (Delmas & Burbano, 2011).

In the built industry, greenwashing may occur in a variety of forms. For green building products such as green roof systems, materials may be labelled as environmentally friendly, sustainably sourced or manufactured, etc. One particularly common claim is that products have a low carbon footprint. While that may be true, data on the actual life cycle cost of the product, including storage and transport to site, may have been omitted. If such data were to be included, the overall carbon footprint of the product may be much higher.

Another common form of product greenwashing is for business entities to claim that greenery can provide multiple benefits (everything listed in Fig. 3) and then to have the greenery product next to these claims. Viewers may unwittingly associate the listed benefits with the product. This is untrue, as species, form and placement are integral to actuating the specific benefits of greenery. For instance, trees can sequester carbon by storing them in their biomass, something green walls cannot do well, as the plants used tend to be small shrubs.

Greenwashing can also occur in the design process, where illustrative renderings may portray lush greenery with ample fauna to conjure images of abundant biodiversity and close approximation to nature. In reality, the actual ecosystem services gained by the design scheme may be limited and overshadowed by disservices that may occur.

In spite of the growing body of literature on ecosystem disservices, a majority of the public still views greenery as a net positive attribute for cities. This is evident from the growing list of countries and cities adopting greenery into their urban plans and building regulations (Bush, Ashley, Foster, & Hall, 2021; Esmail et al., 2022; Tan, Wang, & Sia, 2013). However, without proper planning and execution, it is reasonable to expect greenery to play a role in greenwashing in the built industry. It is a tangible treatment, highly visible and most closely associated with nature and sustainability.

Using greenery as a tool for greenwashing may result in undesirable consequences. When greenery is added for extracting ecosystem services, but without a comprehensive understanding of the mechanism behind these services, they may not be realised to their full potential. Worse still, when a failure occurs (e.g., plants wither on green walls), it may reduce public confidence in the benefits of having urban greenery in the first place, discouraging future attempts at greening the urban environment.

It is, therefore, important to conduct the process of adding greenery in a logical and systematic manner. This book aims to provide a glimpse into urban greenery at different scales and the considerations associated so as to maximise ecosystems services, minimise disservices and avoid green washing, be it unintentional or otherwise.

CHAPTER 3

City-scale greenery

This chapter looks into greenery at the largest scale. Changes implemented at this scale can exert widespread impact to its surroundings, through the apparatus of policy and governance. Sufficiently large areas of greenery can affect climatic conditions as well as spatial structure and dynamic properties of animal migration patterns.

There are quantification methods for greenery unique to this scale, such as via remote sensing techniques and employing landscape metrics. In general, the emphasis here is more on using maps and databases for analysis and less on species and systems-level information. This scale is relevant to the urban, peri-urban and rural zones of the study area (Fig. 2). Intervention for city-scale greenery can be further broken down into city, township, precinct and building levels. Some common intervention strategies at this scale include urban planning masterplans, legislations, government policies, initiatives from organisations, design checklists and green building rating tools (Fig. 4). These methods serve to connect the intervention strategy to ecosystem services and have been used in many cities around the world as part of their urban planning and design process. However, as policies are often simplified for ease-of-use, they tend not to be able to address the complexities or the urban mosaic and inherent heterogeneity of the built environment.

A clearer understanding of the ecosystem services and their spatial attributes, requirements and interactions with other systems is essential for effectively incorporating them into the planning and design process (Troy & Wilson, 2006). This can be challenging to define, as benefits can be experienced directly at the source of green infrastructure or indirectly as a spill-over from adjacent areas. Indirect services may also be affected by external factors such as topography and built infrastructure. Therefore,

City	Masterplans
	Urban planning legislations
Township	Building legislations
Precinct	Policies and Initiatives
	Sustainable design checklists
Building	Green building rating tools

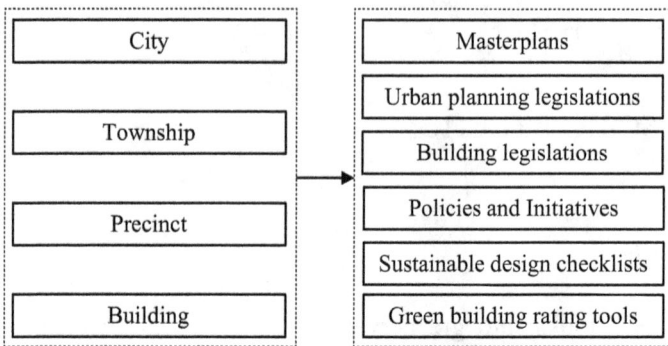

Fig. 4.   Common ways of promoting green infrastructure, from city to building.

it is important to reduce further fragmentation of green spaces during urban development and, where possible, to connect green infrastructure from the inner urban areas all the way to the rural hinterland (Gómez-Baggethun et al., 2013).

Two broad approaches for promoting green infrastructure for the purpose of generating ecosystem services can be considered (Forman, 1995); the former involves strict conservation within a bounded area. Such areas are typically large and are protected from urban development as well as anthropogenic interference. The latter looks into the enhancement of all parts of the existing landscape, emphasising on maximising the potential of the smaller green spaces (Fahrig et al., 2011). Both approaches are equally important in addressing different aspects of ecosystem services.

To drive urban greening initiatives, marketing and publicity campaigns are initiated to garner support and inform the public domain of the underlying objectives. Slogans, catchphrases and illustrative examples are used to effectively convey complex scientific concepts to a wider audience in a more palatable manner. This includes city dwellers as well as practitioners in the built industry, who have to make decisions that are in line with the policies of the area. The addition of greenery itself is often not the end goal but a means to achieve objectives such as increasing the Green Infrastructure (GI) of the city or adopting the Nature-Based Solutions (NBS) approach for developing or redeveloping the city. Some examples are presented in the following sections.

Urban greening policies

Many cities around the world have adopted sustainable development initiatives to address the adverse effects of urbanisation on the environment. Aspects of urban greening can be found in many of these initiatives. They are highly contextual and adopt practical considerations of available resources as well as conditions of neighbouring regions.

The observable framework for urban greening policy development is shown in Fig. 5. The agenda for wanting to introduce the policy is first defined. This can vary from city to city, ranging from mitigating and adaption to climate change, to preserving biodiversity, flood alleviation,

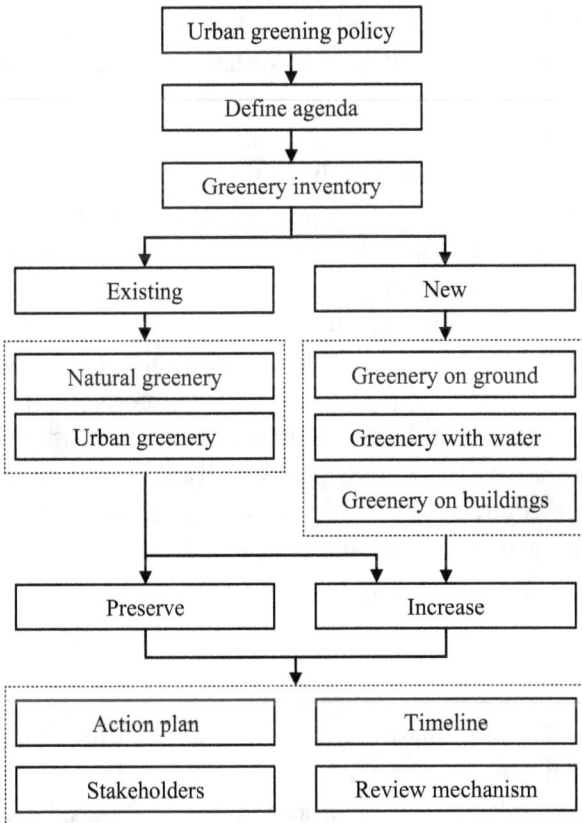

Fig. 5.   Framework for urban greening policy development.

pollution abatement, shoreline protection, aesthetics, raising the quality of living, etc. Subsequently, efforts are made to persevere and increase greenery inventory. This can be categorised into existing greenery, such as forests or urban greenery, like parks and gardens. Thereafter, new green spaces are added. A common strategy adopted by many cities is to begin with the preservation and connection of existing green areas, such as forests and parks, followed by the addition of green spaces and skyrise greenery. Intervention may be in the form of action plans, masterplans, policies, by-laws, campaigns, etc. Broadly speaking, they can be categorised into forested, urban, or park-based action plans. Further delineation between stakeholders such as governmental bodies, corporate entities and community are useful to ensure sustained, active participation. Where possible, roles and responsibilities should be clearly stated for all parties involved. Awareness campaigns are a vital part of the initiatives, as the promotion of greenery is complementary to other sustainability agendas such as recycling, reduction of carbon footprint, biodiversity preservation and energy savings. In this regard, it is crucial to raise awareness and generate buzz through marketing materials such as slogans, social media platforms, websites and tree-planting campaigns, as well as promoting relevant tools to help the industry and community understand the impact of their involvement to green the city.

The timeline of implementation can range from years to a few decades. This is not surprising, considering the sizable resources, time and effort required for landscaping and construction projects. Long-term projects can be broken down into smaller initiatives, each with its own milestones as a means of tracking progress. Targets can be periodically reviewed, but enforcement is rare. Follow-up action is highly recommended, as targets are long-term in nature, and periodic assessment is vital to ensure that cities are on track to achieve them.

Europe

Many regions in Europe have had a long history of incorporating greenery into their town and city planning process (Feng & Tan, 2017). Greenery has played a prominent role in European urban landscapes as early as the 16th century and has influenced the design and planning of cities from other regions (Forrest & Konijnendijk, 2005). The Garden City Movement, for example, started in the early 20th century in England and eventually spread to many parts of Europe and around the world. The role of greenery in the built environment has increased over time, to address issues relating to sustainable development and environmental protection amidst rapid urbanisation.

Cities with high standards of living, such as Vienna, Oslo and Zürich have urban greening policies that not only seek to protect and increase green cover but also improve social capital and well-being of their inhabitants by incorporating campaigns to foster better social interaction through greenery-related activities (Jiang, Menz, & Peric, 2023; Mocca, Friesenecker, & Kazepov, 2020; Røe, 2016). In Berlin, where promoting green cover has been ongoing with policies such as the "Berlin Strategy: Urban Development Concept 2030" and "City Tree Campaign" (Thierfelder & Kabisch, 2016), signs of active citizenry are also evident: citizen-led initiatives such as the "Gieß den Kiez" utilises online platforms to monitor irrigation for street trees as well as to crowdsource manpower to help water and care for them during periods of low rainfall (CityLAB, 2021). In Amsterdam, the "Amsterdam Green Infrastructure Vision 2050" promotes the incorporation of greenery in the city through extensive greening. Community engagement efforts, such as the promotion of allotment parks and sports parks, have been initiated by the city. This is translated into actionable items such as updates to building legislation and standards (de Vries, 2021; Department of Planning and Sustainability, 2021).

Tools such as the Urban Greening Factor (UGF) calculator, used to quantify the amount of greenery provision for new developments in the city of London, serve to provide a more quantifiable means of greening the built environment (Grant, 2017). Similarly, the city of Helsinki has the Green Factor tool, which looks into the provision of green infrastructure for urban planning through the implementation of up to 43 types of green elements (City of Helsinki Environment Centre, 2016; Juhola, 2018).

America

Similar to Europe, cities with high standards of living in America tend to have policies to safeguard greenery. The "Greenest City 2020 Action

Plan", adopted by the city of Vancouver, aims to provide sufficient greenery to ensure that residents can be within a five-minute walking distance from the nearest green space. The "Climate Emergency Action Plan" focuses on the enhancement and protection of natural systems such as forests and shorelines (Rahoui, 2021; Vancouver, 2020). In the city of Toronto, the "Strategic Forest Management Plan" aims to increase canopy cover to 40% by protecting existing green areas and increasing the urban forest footprint (Forestry and Recreation Division, 2013). The "Parkland Strategy" aims to increase the accessibility of parks to city inhabitants within walking distance. Park provision is estimated using the Park Catchment Tool, where the park area is divided by the user population within a walkable distance (500 m) (Parks Forestry and Recreation City Planning, 2019).

Increasing green coverage is a common priority among cities. The city of New York, for instance, has embarked on numerous initiatives to improve living conditions, such as "PlaNYC" in 2007 to improve sustainability efforts, the "Green Infrastructure Plan" in 2010 to address hydrological concerns such as flood and improve water quality, and "MillionTreesNYC" and its goal of planting one million trees in the city (McPhearson et al., 2017). This target was reached in 2015 through collaboration between private, public, as well as non-profit organisations, involving around 50,000 volunteers (Keeley, Benton-Short, Keeley, & Benton-Short, 2019).

Some cities use greenery as a means to improve the overall liveability of the city. For instance, the Detroit Greenways Initiative focuses on making the city more pedestrian and bicycle friendly by creating new trails and recreational spaces infused with greenery. An online map provides useful information on available trails and amenities within the greenways (Detroit Greenways Coalition, 2007; Hartig, Scott, Gell, & Berk, 2022).

The Duwamish Valley, located in Seattle, has long suffered from industrial pollution. To address this issue and other environment-related problems, the "Duwamish Valley Action Plan" proposes a series of interventions, one of which involves parks and open spaces (Calloway, 2019). There has been a strong emphasis on equitability in the planning ethos: that people from all walks of life and socio-economic status are able to have access to parks and can enjoy high-quality spaces for recreation and leisure. Community-led initiatives such as the "South Park Green Space

Vision" help to facilitate project implementation and community engagement (Seattle Parks Foundation, 2014).

The city of Curitiba in Brazil is one of the greenest cities in the world. Awarded the "most ecological city" by the UN in 1992, Curitiba has incorporated numerous environmental programs in its urban planning process. In the 1970s, Curitiba's "Green Plan" focused on addressing the issue of flooding in the city. The "Biocity Program", launched in 2007, looked into the preservation of biodiversity amidst development in urban areas through various projects. The "Live Barigui Project", for instance, aimed to improve hydrological conditions in the area by creating an ecological corridor that serves to improve water quality and increase local flora and fauna count. This is further supported by policies that ensure the protection of forested areas, as well as incentives such as tax breaks for privately owned estates that preserve large areas of green space (Zingoni de Baro, 2022).

Australasia

Similar to Europe and America, major cities in the Australasia region also recognise the need to preserve the urban forest and that all forms of greenery can help improve conditions in the urban environment. Some cities devote more attention to their indigenous heritage and culture, seeing it very much as a vital part of their natural capital. For instance, the Auckland Council in New Zealand stresses on the cultural significance of preserving their urban forest, with reminders of their Māori heritage and close association with nature (Auckland Council, 2019):

> "...The urban ngahere is an important part of Tāmaki Makaurau/ Auckland's cultural heritage. Remnants of native forest represent traditional supermarkets (kai o te ngahere), learning centres (wānanga o te ngahere), the medicine cabinet (kapata rongoā), schools (kura o te ngahere) and spiritual domain (wairua o te ngahere). Trees also represent landing places of waka (canoe) and birth whenua (to Māori, it is customary to bury the whenua or placenta in the earth, returning it to the land)..."

> — Excerpt from Auckland's "Urban Ngahere (Forest) Strategy"

This forms the underlying philosophy of the "Urban Ngahere (Forest) Strategy", the urban greening plan adopted by Auckland, partly to address the impacts of urbanisation and changes to tree protection policies in the area. A strategic framework outlines how to increase green cover and to optimise the positive impacts of greenery, such as having the right tree at the right place and creating ecological corridors for biodiversity preservation. It is part of the "Auckland Plan 2050" to ensure sustainable long-term development for the city (Auckland Council, 2018). Similar to "MillionTreesNYC", the "1 million trees programme" launched in 2017 aims to increase green cover by planting a million trees and shrubs. Besides beautifying the city, the initiative also aimed to protect waterways and harbours by reducing siltation. Implementation involved not just the government but also corporate partners as well as schools and communities.

In Wellington (New Zealand), the "Our Natural Capital — Wellington's Biodiversity Strategy & Action Plan" adopted in 2015 aims to preserve and restore indigenous biodiversity through a slew of policies and management plans, including the creation of connections between green areas for flora and fauna via a "Green Network Plan" (Wellington City Council, 2015). The action plan focuses heavily on biodiversity and is executed in acknowledgement of the Māori approach to managing resources and respecting the symbiotic relationship between living organisms and their ecosystems.

In Australia, the city of Melbourne pays much attention to its green infrastructure, with policies such as the "Green Our City Strategic Action Plan" as well as the "Urban Forest Strategy" providing comprehensive greening strategies for the city in terms of both skyrise greenery as well as ground level green cover (Melbourne, 2012; 2017). The city also has its own green infrastructure assessment tool. Named the "Green Factor Tool", it is an online tool that can be used by practitioners to measure the efficacy of green infrastructure for their development projects. This tool is site-specific to the city of Melbourne and cannot be used for cities outside of the territory. It should also be noted that the ecosystem service benefits are prioritised according to conditions in Melbourne and may not be well-suited if applied to other cities directly. For instance, air purification as an ecosystem service benefit provided by greenery was not considered a priority as cities in Australia did not suffer from poor urban air quality, compared to cities from other parts of the world. Additionally, policies that directly targeted fossil fuel emissions were deemed to be more affected than greenery policies in this area (Bush et al., 2021).

As part of the "Sustainable Sydney 2050" initiative, the "Greening Sydney Strategy" seeks to tap on various ecosystem services of urban greenery in the development of the city of Sydney. Targets for action plans are based on feedback from community engagement sessions, with reviews conducted every ten years (Sydney, 2021).

Asia

Many cities in Asia have also included greening practices in their urban development policies. In China, national agendas help provide direction and consistency at the provincial, prefectural and county levels. The establishment of an "ecological civilisation" axiom focuses on environmental protection and sustainable development, in addition to economic growth and prosperity for the country (Xie, Flynn, Tan-Mullins, & Cheshmehzangi, 2019).

There have been several national-level greening initiatives to increase green cover for cities in China. The "National Garden City" title is awarded to cities that have achieved high standards of the sustainable urban environment and greenery provision (Wu & Kim, 2021). Severe flooding events in recent years have brought about increased urgency to improve stormwater management through nature-based solutions. The "National Sponge City Programme" aims to significantly increase blue-green infrastructure in China. The pilot programme consisted of 50 Chinese cities and involved engineering projects such as improving drainage networks and building swales and parks to improve flood management (Fu, Zhang, Hall, & Butler, 2023). Such national programmes have spurred further initiatives to increase green cover at the provincial and city level, leading to the creation of local policies and guidelines to aid in sustainable urban planning practices. The "Pearl River Delta Greenway", for instance, aimed to increase the proportion of green paths in the region. These areas served as green amenities for recreational use, such as cycling and exercise. In 2017, the Chengdu municipal government launched the "Greenways of Paradise" programme to increase the number of greenways amidst rapid urbanisation. Ecological zones are set up within the greenways to limit urban development and to preserve the ecological quality of these areas (Zhang, Wu, & Lin, 2022). The "Beijing City Master Plan (2016–2035)" provides direction to increase the number of urban forests as well as pocket parks in the city of Beijing (The People's Government of Beijing Municipality, 2020). The urban greening effort takes into consideration major man-made and natural landmarks, adopting a "one barrier, three rings, five rivers, nine wedges" ecological planning principle to include the mountainous region, green rings, rivers as well as ecological areas connecting the central part of the city to its periphery (一屏、三环、五河、九楔) (Zhang, Zhang, Zeng, Yu, & Zheng, 2021).

The "Hong Kong 2030+" plan provides a framework for urban development in Hong Kong for the future, with emphasis on increasing blue-green infrastructure (Planning Department, 2021). Besides the framework, various building standards and guidelines help to ensure sufficient green coverage for the city. The Hong Kong Planning Standards and Guidelines (Chapter 4: Recreation, Open Space and Greening) provides useful information on greening the city, with references to more detailed, technical information for practitioners (Planning Department, 2015). For instance, the "Greening Master Plan" outlines unique themes for the major urban regions of Hong Kong: Tsim Sha Tsui adopts the theme of "Jade Necklace" and the Western District has the theme of "Civil Elegance and Joyful Renaissance", playing off unique urban characteristics found in the respective areas (Civil Engineering and Development Department, 2019).

In Seoul, policies relating to green spaces and parks in the metropolis can be found in the "2030 Seoul Parks and Green Space Master Plan", which aims to increase green infrastructure to address the adverse effects of urbanisation such as the UHI effect and air and water pollution (Park, Yang, & Jang, 2012). Over the years, this has led to many urban regeneration projects, such as the Cheonggyecheon stream restoration and the "A Thousand Forests, A Thousand Gardens" initiative to plant 30 million trees in the city (Choi & Kim, 2022).

Southeast Asia

Cities in Southeast Asia are experiencing rapid urban expansion, with a projected population of 373 million by 2030 (Khuong, McKenna,

& Fichtner, 2019). The tropical hot and humid climate has led to a high reliance on mechanical cooling (HVAC), resulting in increased energy consumption and climate-related problems such as the UHI effect (Santamouris, 2015). As such, cities in Southeast Asian countries tend to focus on policies to reduce energy consumption via active and passive means, as well as to improve outdoor thermal comfort. There is also an emphasis on addressing seasonal flooding in cities by introducing blue-green infrastructure.

In Malaysia, policies such as the "Kuala Lumpur Climate Action Plan 2050" advocate for the reduction of carbon footprint through various initiatives. Under the "Green Adaptive City" strategic roadmap, incremental steps are taken to increase green cover, reduce impermeable surfaces and promote low-impact development (KL City Hall, 2021).

In Vietnam, the "Hanoi Master Plan 2030", established in 2011, aims to promote sustainable urban development by introducing green spaces such as green belts and buffers in strategic areas (Wilson, 2009).

The city of Bangkok has been involved in urban greening policies through tree planting campaigns since the 1990s (Thaiutsa, Puangchit, Kjelgren, & Arunpraparut, 2008). An urban green space policy of providing 5 m^2 of green space per capita was initiated in 2007. With multiple policies supporting the addition of green infrastructure in the city, green space provision reached 6.71 m^2 per capita in 2019. The Bangkok Metropolitan Administration (BMA) continues to promote green space expansion via policies and legislation such as the "Thailand's National Strategy" and "Town Planning Act B.E. 2562" (Chanchitpricha & Fischer, 2022).

To capitalise on ecosystem services provided by greenery, the "Spatial Plan Jakarta 2030" masterplan includes a "Green Open Space Plan" that aims to maintain 30% of the city of Jakarta as green open space (20% for public spaces and 10% for private spaces) (Setiowati, Hasibuan, & Koestoer, 2018). The Jakarta Green Factor, similar to the Biotope Area Factor, can be used to estimate the ecological value of a site by measuring its green cover (Dahlan, Faisal, Chaeriyah, Hutriani, & Amelia, 2022).

Greenery provision has been a prominent feature of urban planning and development in Singapore since the formation of the city-state in 1965. Over the years, the role of greenery has evolved from being a largely cosmetic treatment to one that seeks to address pressing issues of environmental sustainability and liveability in the urban landscape (Tan, Wang, & Sia, 2013). Faced with spatial constraints and an ever-expanding population, the approach to greenery provision includes both the preservation of existing woodland as well as policies to ensure sufficient greenery coverage for built-up areas. The Urban Redevelopment Authority (URA) oversees the provision of greenery in parts of the city-state. The "Landscaping for Urban Spaces and High-Rises" (LUSH) is a scheme that consolidates urban greening efforts through building requirements and incentives. Under this scheme, new developments in designated zones are required to provide greenery replacement that is equivalent to the gross site area (100% replacement). Different forms of greenery, such as planter boxes, green roofs and green walls, qualify as Landscape Replacement Areas (LRAs). Incentives such as Gross Floor Area (GFA) exemption are given for projects that fulfil stipulated LRA criteria (URA, 2017). The Skyrise Greenery Incentive Scheme, introduced by the National Parks Board in 2009, funds up to 50% of installation costs of vertical and rooftop greenery (Beatley, 2016). To ensure consistent quantitative description of greenery across agencies, the Green Plot Ratio (GnPR) is the common metric used to calculate green cover for multiple initiatives such as the URA LUSH scheme, the Building Construction Authority (BCA) Green Mark scheme, and Housing & Development Board (HDB) design guidelines for new residential housing precincts (Schröpfer & Menz, 2019). Besides governmental initiatives to increase green cover, there have been continuous efforts to engage the wider public to green up the city with aspirant visions. From "Garden City" (1967) to "City in a Garden" (2012) to "City in Nature" (2023), a key focus area in the "Singapore Green Plan 2030" is to create sustainable and liveable homes for its inhabitants. By 2030, it aims to increase tree-planting efforts and increase park space such that every household is within a 10-minute walk from a park (Singapore Green Plan, 2021).

Green building rating tools

In addition to urban planning policies, greenery provision can also be found in green building rating tools. Here the term *"green"* does not explicitly refer to *"vegetation"* but alludes to concepts of *"sustainability"* and *"environmentally friendliness"*. The practice of green building is one that emphasises the efficient use of resources such as materials, energy and water, with minimal impact on the environment and its inhabitants. This is achieved through a combination of both active and passive strategies.

The green building movement dates back to the 1970s, when the built construction industry started to be perceived as a major source of resource consumption and environmental degradation. This led to a push to relook traditional building practices and improve energy efficiency (Steele, 1997). A call for more sustainable measures of building construction led to the birth of the green building movement, as well as a need to develop a standard to verify if a building was indeed "green". This led to the creation of the green building rating tool, which functions as a building appraisal checklist for a building's green building features (Ade & Rehm, 2020). Using the green building rating tool, designers can strive for sustainable design solutions without drastic infringement of creativity and design freedom.

Green building rating tools are predominantly used to assess individual buildings, but there are also specialised rating tools for precincts, neighbourhoods, as well as different building typologies (Szibbo, 2016). Besides energy efficiency and resource conservation, some checklists also focus on aspects of human well-being (Licina & Yildirim, 2021). Greenery is mentioned in many rating tools, under sections such as habitat restoration, environmental protection and land use and ecology. Some (non-exhaustive) examples are provided in the following section.

BREEAM

The Building Research Establishment Environmental Assessment Method (BREEAM) is a green building assessment method published by the Building Research Establishment (BRE) in 1990. It can be considered to be the first green building rating system for residential and office buildings (Yudelson & Meyer, 2013). Widely adopted around the world, BREEAM has been used in over 86 countries and for more than 570,000 buildings (BRE, 2020).

An example of where greenery in BREEAM is mentioned is in the Asset Performance and Management Performance criteria (BRE, 2020).

In Asset Performance, greenery can be found in the "Resilience" and "Land Use and Ecology" categories. Under "Resilience", green roofs are listed as possible solutions to mitigate surface water runoff impact. Under "Land Use and Ecology", credits are given for planted areas and specific ecological features of the planted areas. For the purposes of ecology and improving human well-being, planting areas have to occupy between 5% and 70% of the building footprint. This can include vertical greenery. The formula for calculating the percentage of the planting area is shown here:

$$Ap = \frac{\sum (Ae + Av)}{Af} \times 100, \tag{1}$$

where

Ap = Percentage planting area (%)
Ae = External planting area (m²)
Av = Vertical planting area (m²)
Af = Asset footprint (m²)

For Management Performance, greenery can be found in the "Land Use and Ecology" category. Credit is given for having a biodiversity management plan. Landscaping works must be aligned to the proposed biodiversity management plan. The ecological value of the asset has to be assessed periodically, and the management plan has to be updated depending on the ecological state of the project.

LEED

Leadership in Energy and Environmental Design (LEED) is a green building certification tool managed by the US Green Building Council since 1998. Buildings certified under LEED earn points by addressing issues relating to energy, water, materials, environmental quality, etc. There are different rating systems offered to various types of projects and buildings, such as retail, healthcare and neighbourhoods (Champagne & Aktas, 2016). There are 100,000 LEED-certified projects in the world, ranging from residential projects to hospitals, schools, stadiums and offices (Homes, 2022).

An example of where greenery in LEED v4.1 is mentioned is in the "Sustainable sites" and "Water efficiency" areas of focus (USGBC, 2022). Under "Sustainable sites", there is a requirement of planting at least six native or adapted species of plants for the purposes of habitat protection and restoration. Native flowering plants must be specified for the pollinator garden, and the area should be at least 3 m². This is under the "Protect and Restore Habitat" credit.

Vegetation is also mentioned in the "Open Spaces" credit. At least 25% of outdoor spaces should be allocated for greenery or comprise vegetated canopy cover. This includes green roofs that are accessible to all.

Under the "Heat Island Reduction" credit, greenery is mentioned as a non-roofing method to reduce heat gain through shade provision, as well as vegetated roofs comprising native or adapted plants.

Under the "Places of Respite" credit, outdoor spaces in healthcare facilities are recommended to be planted with multiple species of native or adapted plants. Selection of plant species for reducing irrigation for landscapes is mentioned in the "Water efficiency" area of focus.

WELL

WELL is a building certification program that, unlike BREEAM and LEED, places more emphasis on efficient energy and water usage and focuses on the impact of building attributes on human health and well-being. Managed by the International WELL Building Institute, WELL looks into the areas of Water, Lighting, Air, Movement, Sound, Thermal Comfort, Mind, Innovation, Nourishment, Materials and Community. Greenery can be found in the areas of Nourishment, Thermal Comfort and Mind (WELL, 2022).

Under the Nourishment category, the "Food Production" section awards points for providing a gardening space that is permanently accessible and within 0.4 km of the building project. The space has to include either an edible landscaping, aeroponic or hydroponic system. Food grown from the gardening space should be made available to building occupants.

Under the "Outdoor Thermal Comfort" section of the Thermal Comfort category, 1 point is given for having outdoor areas shaded by tree canopies. A total of 50% of building entrances and pedestrian pathways should be covered, followed by 25% to 75% coverage for other outdoor areas. Alternatively, simulation can be done to ascertain the impact of the proposed intervention on human thermal comfort.

Under the Mind category, the "Nature and Place" section encourages the addition of natural features such as plants to connect occupants with nature. In the "Restorative Spaces" section, a point is given for providing a restorative space for relieving burnout and stress. This space should have a smoothing ambience and include nature or natural elements, such as plants. In the "Enhanced Access to Nature" section, emphasis is given to featuring nature and natural features, both inside and outside the building. Points are given for views to indoor planting, as well as access to outdoor spaces that are landscaped with lush greenery.

BCA Green Mark Scheme

The BCA Green Mark Scheme is a green building certification program used for new and existing buildings in Singapore. Launched in 2005, it aims to increase the stock of green buildings in Singapore by assessing the categories of Energy Efficiency, Resilience, Whole of Life Carbon, Health and Well-being, Intelligence and Maintainability (BCA, 2022c).

Greenery provision is a mandatory requirement for all tiers of Green Mark rated buildings and can be found in the Health and Well-being as well as the Resilience categories.

Under the Health and Well-being category, "Access to Nature" section, between 2.5 and 3 points can be earned by providing spaces such as sky terraces and roof gardens that are accessible to building occupants. Common areas should contain natural features such as plants. Visual access to greenery should be provided if it is not practical to have indoor green spaces. Under the "Restorative and Community Spaces" section, communal/sensory gardens should be provided within the building parameters for facilitating restorative activities (BCA, 2022a).

Under the Resilience category, the "Habitat and Ecology" section outlines the usage of the Environmental Impact Assessment (EIA) for construction projects to identify potential environmental impacts due to construction and operational activities as well as possible mitigation methods. Greenery is not explicitly mentioned here but is closely related to EIA topics such as landscape ecology and biodiversity preservation. Under "Urban Heat Island Mitigation", points are given for areas that show UHI mitigation measures, such as the provision of green spaces and rooftop greenery to reduce temperature. Under the "Restore" section, points are awarded for greenery provision (Green plot ratio of more than 5 for new buildings). Provision of native plants as well as "wild landscape areas" are encouraged. Under "Natural Climate Solutions", restoration of ecology equivalent to the GFA of the construction project via initiatives such as reforestation programmes provides up to 2 points (BCA, 2022b).

Measurement

Urban greenery are value-creating systems, but they also consume resources. The lack of proper inventory and management may result in unexpected consequences. At best, the expected ecosystem service benefits may not be fully realised. At worst, greenery may end up expending more resources than the benefits they provide. The net deficit can reduce public confidence in the benefits of having greenery in the urban environment and contribute to greenwashing.

Measurement of greenery is therefore useful for the following reasons:

1. To detect trends and patterns of green coverage over time. Metrics such as the Park Provision Ratio (hectare per 1,000 persons) can be used to conveniently describe greenery provision at the city scale and uncover longitudinal trends as well as review performance targets (Tan, 2017).

2. To promote greenery provision or design towards a certain agenda, such as to tap on specific ecosystem services. To realise the benefits brought forth by greenery (e.g., temperature reduction), both quantity as well as quality of greenery must be sufficient. A clear framework for describing the amount and type of greenery provided and its associated benefits is necessary to encourage confidence and widespread adoption.

3. To assess landscape performance based on a set of pre-defined criteria. Green spaces can be, to an extent, objectively appraised for their potential to improve the urban environment. Some common criteria include promoting biodiversity and enhancing air quality. The expected benefits tend to be extracted from reviewed literature and have some correlation to landscape composition and configuration.

There are numerous ways of measuring greenery, and the exact method employed should be complementary to the objective(s) of the measurement exercise. In addition to greenery composition and configuration, the two other commonly measured items are anthropocentric and ecological quality. These metrics serve to help users estimate corresponding human and ecological impacts. The four main areas of measurement are as follows:

1. Greenery quantity. This scope of measurement focuses on the amount and type of greenery that is present in the surveyed area. The variety and abundance of different forms of greenery, such as the size of parks, forests, etc. Also known as landscape composition.

2. Greenery quality. This goes a step further to evaluate more distinctive parameters of green spaces, such as their shape, orientation, edge conditions, spatial distribution, etc. Commonly termed as landscape configuration metrics.

3. Ecological quality. The ecological processes found within a landscape are a function of its spatial attributes as well as the composition and configuration of its elements. Some examples include the level of biodiversity, water quality, nutrient cycling, etc. Measurements of ecological quality provide some indication of the status of ecosystem services provided by the landscape in question.

4. Anthropocentric quality. This scope of measurement is concerned with the potential benefits for humans due to the presence of greenery (Items 1 and 2). In this case, greenery is associated with natural processes or functions that can improve quality of life (supported by literature). Some examples of ecosystem services include the improvement of outdoor thermal comfort and increase in visual access to greenery.

When deciding on how best to measure greenery, it is important to first be clear on the objective for wanting to do so. Strictly speaking, the intention is not so much to "measure" greenery but to describe greenery in a manner that can facilitate analysis and decision-making. Depending on the chosen method, the description can be quantitative or qualitative in nature. In some cases, it can be a mixture of both.

Some key questions to keep in mind include:

1. What is the purpose of wanting to describe greenery?
2. What aspects of greenery does the selected method look into?
3. What is the expected outcome of using this method?

It is useful to consider the following framework when deciding on the appropriate measurement metric (Fig. 6). The examples listed below provide an overview of some ways greenery can be described and are non-exhaustive.

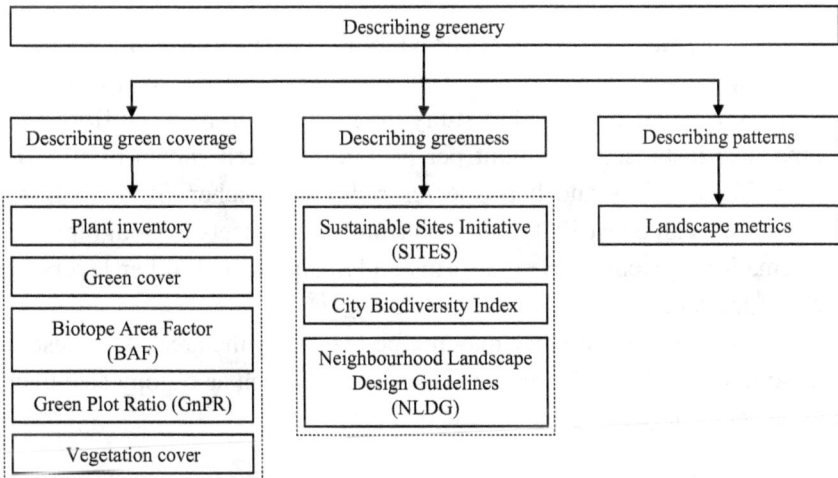

Fig. 6.   Framework for describing greenery.

Describing green coverage

Green coverage can described in a quantitative manner. This typically involves the aggregation of vegetation data such as plant count, green area, total leaf area, etc., over a surveyed area.

Plant inventory

One direct method of describing green coverage is to tally all the plants of interest. This provides basic information on the plant inventory for a given area and can be normalised for comparison between different sites (Equation (2)).

$$Plant\ inventory = \frac{\sum P}{A}, \tag{2}$$

where

P = Plants of interest

A = Area

Example: A site of 100 m² has 40 trees and 10 shrubs.

It has 40/100 = 0.4 trees/m² and (40 + 10)/100 = 0.5 plants/m²

Such metrics are useful to provide a quick overview of greenery inventory as well as for comparison over time and between different sites. They can also be used to study the relationship between flora and fauna. For instance, tree count per hectare has been shown to significantly impact birds and bats species richness (Fischer, Stott, & Law, 2010). Data can be easily transferred from spreadsheets to Geographical Information System (GIS) platforms and analysed with other layers of geospatial data.

Having a plant inventory may not be feasible if the area of interest is too large for each plant to be inventorised. Data such as canopy size may not be adequately captured via this method.

Green cover

When the area of interest is too large to maintain a plant inventory (e.g., township scale), green cover can be used instead. Green cover involves the quantification of greenery via area coverage. It is typically used for large-scale quantification and is commonly used in research and urban planning. For this metric, the focus is less on individual trees and more on large green spaces such as parks and forested areas. Similar to plant inventory, both the absolute value (e.g., area of park space) (Equation (3)) and the normalised data (e.g., park space per 1,000 resident) can be used for analysis.

$$Green\ cover = \sum G, \tag{3}$$

where

G = Green areas of interest

Green cover, or derivatives of green cover metrics, can be used to describe greenery provision at the city scale. Some examples include park provision ratio (total area of park space per capita), park area or forested area over the total area. These metrics help to provide some indication of green cover in the area of interest and can be used for comparison between different areas (Tan, Wang, & Sia, 2013). They can also be used for analysis with other independent variables such as population demographics, socio-economic status, etc.

Note that the term "park space" includes everything within the park's boundary and denotes both hardscape and softscape. There is no clear distinction between the different forms of greenery, such as trees, shrubs and turf. This may become a limitation for studies that require a higher level of granularity when it comes to the description of green spaces.

Biotope Area Factor

Green cover can be used to assess greenery provision. To a certain extent, it can also be used to estimate the ecological value of the area, depending on the type and density of green cover. One metric that incorporates both green cover and ecological potential is the Biotope Area Factor (BAF). The BAF was first introduced in Berlin as a means to improve ecological conditions in urbanised areas and to ensure environmental protection in landscape planning (Landschaft Planen & Bauen, 1990).

A "biotope" is defined as an area with homogenous conditions that is ecologically effective, which can mean a habitat for flora and fauna or the provision of ecosystem services such as runoff retention. Biotopes are ranked according to their characteristics and perceived ecological value. The perceived value is then multiplied by the area and aggregated over the surveyed area to give an overall score (Equation (4)):

$$BAF = \frac{\sum(EEA \times EVF)}{A},\qquad(4)$$

where

BAF = Biotope Area Factor
EEA = Ecologically Effective Area (m²)
EVF = Ecological Value Factor per sqm
A = Site Area (m²)

The ecologically effective area and corresponding value factor is pre-determined (Fig. 7) and weightages can be calibrated according to the local context. Since the inception of the BAF, this metric has been adopted by cities such as Malmo, Seattle and Seoul (Keeley, 2011).

In theory, using the BAF as a planning tool allows for flexibility in the design process. Designers are free to propose solutions without restriction to form and space as long as the stipulated BAF requirements are met. This provides a higher chance for green areas to be developed in urbanised areas. The BAF can also be used for before-and-after comparison scenarios to ensure that development proposals do not adversely impact initial ecological conditions. The BAF can also be applied at the regional scale via remote-sensing methods (Lakes & Kim, 2012). This makes it possible to assess large swaths of land for their ecosystem services in a convenient manner.

Although the potential advantages of using the BAF for precinct or city planning have been widely discussed, it is important to note that, like most greening metrics, the actual impact of using the BAF for planning remains unknown. There is heavy reliance on theoretical frameworks such as ecosystem services and green infrastructure to approximate the potential benefits that a higher BAF score will bring.

While metrics are generic and facilitate the ease of use for practitioners, over-simplification of indicators means that some data will be lost.

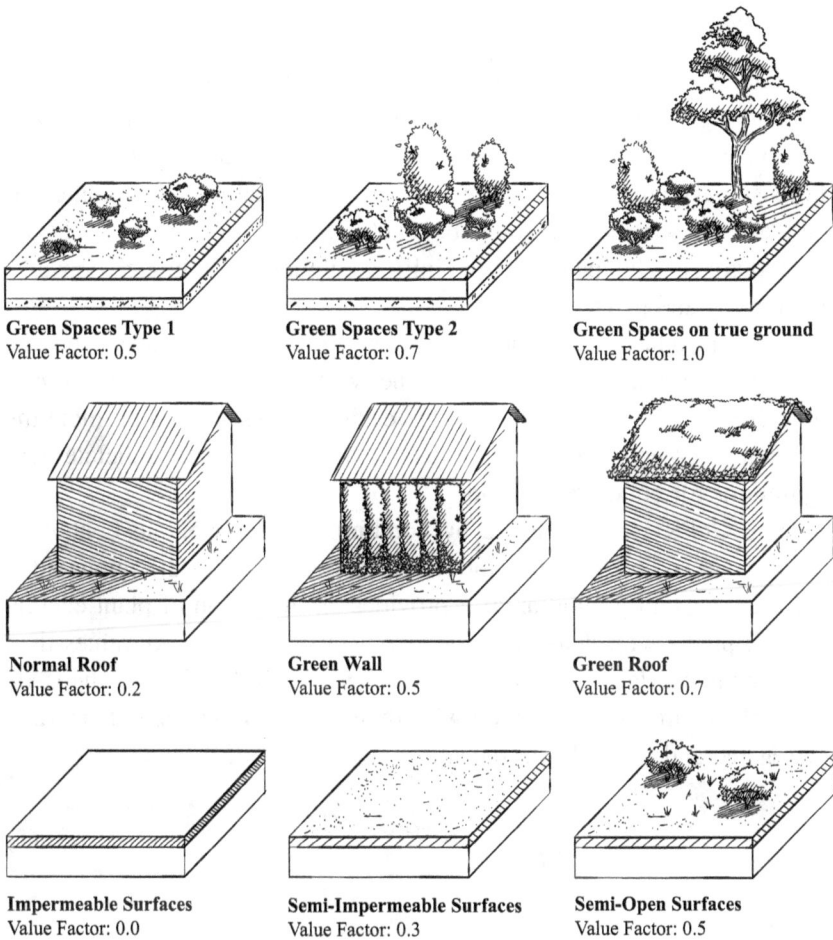

Green Spaces Type 1
Value Factor: 0.5

Green Spaces Type 2
Value Factor: 0.7

Green Spaces on true ground
Value Factor: 1.0

Normal Roof
Value Factor: 0.2

Green Wall
Value Factor: 0.5

Green Roof
Value Factor: 0.7

Impermeable Surfaces
Value Factor: 0.0

Semi-Impermeable Surfaces
Value Factor: 0.3

Semi-Open Surfaces
Value Factor: 0.5

Fig. 7.   Example of an ecological value scoring system. (Adapted from (Keeley, 2011).)

For instance, further differentiation of vegetation and surface types may further increase the variety of weighting ratios, leading to more design choices. However, this may end up confusing users and reduce the appeal of using such metrics as part of the planning and design process.

Also, the weightage assigned to different biotopes for the BAF and other related Green Area Ratio metrics may vary slightly. This may lead to ambiguity in establishing benchmarks and challenges when attempting to compare scores between cities.

Green Plot Ratio

The Green Plot Ratio (GnPR) is a metric that quantitatively describes the amount of greenery in a given site. The key difference between GnPR and previously discussed greenery metrics is that it provides a holistic, three-dimensional quantification of greenery by incorporating both biological as well as site parameters, going beyond land coverage. Technically, all planted area can be included into GnPR tabulation, including green walls, green roofs and planter boxes.

First introduced as a concept by Ong (2003), GnPR has since been applied in numerous studies around the world. It was conceived to integrate architecture and greenery in the design process and to promote the addition of greenery into built projects, without the need to dictate configurations for building and land.

The GnPR was developed by combining aspects of the Leaf Area Index (LAI) and building plot ratio. The LAI is defined as the single-sided leaf area per unit ground area. It provides an indication of plant canopy density: plants with dense canopies would have higher LAI values than plants with less dense canopies. To calculate the GnPR of a site, the LAIs of all plants are first multiplied with their respective canopy areas, then aggregated and divided over the site area (Equation (5)):

$$GnPR = \frac{\sum (LAI_n \times CA_n)}{A},$$ (5)

where

$GnPR$ = Green Plot Ratio
LAI = Leaf Area Index
CA = Canopy area (m^2)
A = Site area (m^2)

The suggested values to use for calculation are shown in Table 2. Species-specific details can be found on Flora & Fauna Web, a free-access online database managed by the National Parks Board of Singapore (https://www.nparks.gov.sg/florafaunaweb).

Since its inception, the GnPR has been used for both practice and research. For instance, the GnPR is used to calculate greenery provision in

Table 2.   Suggested LAI and area values (Tan & Sia, 2010).

	Trees	Palms	Shrubs and groundcovers	Turf	Green wall
LAI	Open canopy = 2.5 Intermediate canopy = 3.0 Dense canopy = 4.0	Solitary = 2.5 Cluster = 4.0	Monocot = 3.5 Dicot = 4.5	2.0	2.0
Area	60 m²	Solitary = 20 m² Cluster = 17 m²	Planted area	Planted area	Covered area

the BCA Green Mark green building certification scheme. It has also been used to study the correlation between greenery and air temperature at the estate level (Wong et al., 2011). Although it is a convenient metric for tabulating green coverage of a building or an estate, GnPR is not able to provide any clear indication of the level of biodiversity or other ecosystem services. The lack of species-specific information and generalisation of plant attributes has drawn criticism over its efficacy in promoting sustainable design solutions. For instance, designers may opt to have greenery in unsuitable locations just to increase the GnPR score (Schröpfer & Menz, 2019).

Vegetation cover

Vegetation cover is a useful indicator when studying a large expanse of land with a need to separate green cover from non-green elements. High-resolution satellite images or aerial photography can be referenced to demarcate green areas, typically with a Geographical Information System (GIS) platform. However, using natural colours (a composite of red, green and blue bands of the electromagnetic spectrum) to identify green areas may pose a challenge if non-vegetated areas have similar colours, such as green-coloured roofs and pavements. This may be resolved by employing remote sensing image analysis methods.

   Solar radiation incidents on live vegetation can be absorbed, reflected or re-emitted, depending on the amount of chlorophyll in the plant. In healthy vegetation, radiation from the blue and red spectrum is absorbed by the chlorophyll, and green is reflected. Much of the near-infrared (NIR) radiation is also reflected, to minimise overheating and tissue damage. Therefore, plants would appear relatively brighter in the NIR spectral region. This difference in reflectance between plant and non-plant objects is used to determine their spatial distribution (Weier & Herring, 2000).

Instead of the red, green and blue band composite for image analysis, wavelengths from the NIR and red spectral region are used instead. The Normalised Difference Vegetation Index (NDVI), which is a mathematical indicator of the greenness in vegetation, is expressed in Equation (6).

$$NDVI = \frac{NIR - Red}{NIR + Red},\tag{6}$$

where

NDVI = Normalised Difference Vegetation Index
NIR = Near Infrared spectral reflectance
Red = Red spectral reflectance

Values for NDVI typically range from –1 to 1 (Table 3). Once the NDVI map is obtained, further classification can be done to further delineate water, non-green and green areas (Fig. 8).

NDVI has been used extensively in environmental studies. Maps generated with the NDVI can be used to identify different land covers and biozones at large scales (Soriano & Paruelo, 1992). Since the NDVI is highly correlated to vegetation productivity, it is a useful index for ecological studies, such as describing the spatial and temporal distribution of plant communities, CO_2 fluxes, hydrology, disaster risk monitoring, drought monitoring and land use change over time (Pettorelli et al., 2005).

For urban planning purposes, the NDVI can be used as an indicator for the provision of ecosystem services such as climate regulation and stormwater runoff in the masterplan (Spahr, Bell, McCray, & Hogue, 2020). As a convenient metric to describe green coverage, the NDVI has become an increasingly popular tool that enables practitioners to adopt a more scientific and evidence-based approach to planning and design. For example, in the ecosystem service modeling exercise for the Tengah Town project in Singapore, the NDVI was used to estimate the effects of greenery on

Table 3. NDVI ranges for different
land cover classes (Akbar et al., 2019).

Class	NDVI range
Water	–0.28–0.015
Built-up	0.015–0.14
Barren land	0.14–0.18
Shrub/grass	0.18–0.27
Sparse vegetation	0.27–0.36
Dense vegetation	0.36–0.74

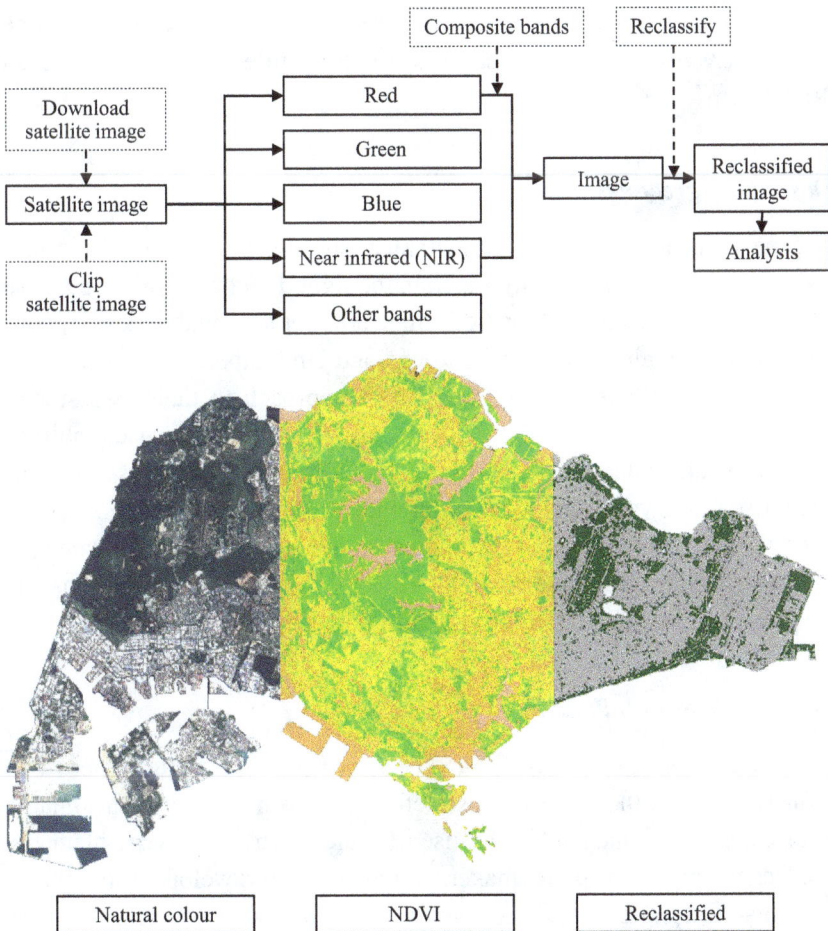

Fig. 8.   Workflow for vegetation cover analysis via NDVI imagery.

variables such as air temperature, runoff retention and soil erosion for the township masterplans (Tan, Gaw, Masoudi, & Richards, 2021).

Although NDVI maps can be used to differentiate different land forms, it is not possible to use it to distinguish between plant species, as different assortments of plants may ending up yielding similar NDVI values (Nagendra, 2001). The fidelity of the NDVI map is also limited by the resolution of the original satellite image. Resolution can vary from 250 m × 250 m per pixel from MODIS, to 30 m × 30 m from Landsat 8, to 10 m × 10 m from Sentinel-2, to 1 m × 1 m from the National Agricultural Imagery Program (NAIP) (Jimenez, Lane, Hutyra, & Fabian, 2022).

A video on how to download, process and visualise the NDVI can be found on www.ugl.sg/book, under the section titled "How to create an NDVI map".

Describing greenness

Greenery in itself does not guarantee sustainable design outcomes, although having it is arguably a step in the right direction. For the purpose of moving towards more holistic frameworks for sustainable design, construction and maintenance of buildings and landscapes, some evaluation systems have adopted a more qualitative approach to their assessment, with an emphasis on describing the greenness (i.e., level of sustainability) for a project. Instead of quantifying greenery provision, these systems comprise frameworks that focus on first identifying the specific ecosystem services to tap on. Suggestions for intervention are only put forth when the theoretical objectives have been identified. A wide range of solutions can be proposed, and they are not limited to greenery provision.

Sustainable Sites Initiative

The Sustainable Sites Initiative (SITES) comprises a rating system that aims to promote the design, construction and maintenance of sustainable landscapes. It is intended to be used by architects, designers, planners, landscape architects, policymakers, engineers and developers to achieve sustainable design, protect ecosystems, and tap on ecosystem services.

Projects can be SITES-certified by achieving a minimum amount of points from the checklist. Certification is voluntary.

The SITES rating system does not measure greenery explicitly but instead assesses the level of sustainability achieved by a project in ten categories. For every category, there is a set of pre-requisites to be achieved, followed by credits that can be earned. Certification is achieved after meeting all prerequisites and obtaining sufficient credits (Fig. 9). Certification consists of four tiers, starting from SITES Certified (70–84 points), SITES Silver (85–99 points), SITES Gold (100–134 points), to SITES Platinum (>135 points) (SITES, 2022).

There is no maximum size of a SITES project, but the recommended minimum size is 185.8 m^2. This may make the rating tool suitable only for precinct-level developments and above.

Since the SITES rating system aims to provide a holistic assessment of the development project, it takes into consideration site-specific indicators and factors that tap on the concept of ecosystem services and sustainable practices not just from greenery but also from aspects of human

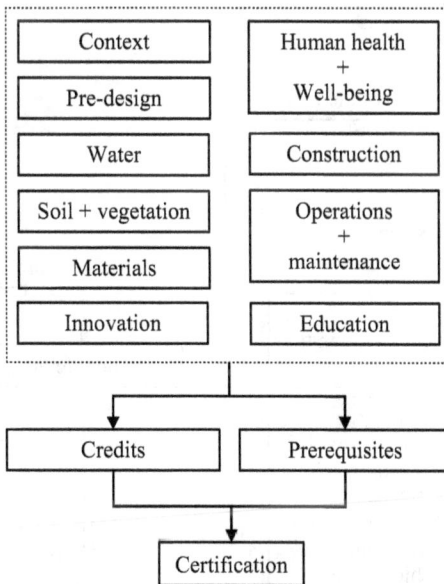

Fig. 9.   Certification framework for the SITES rating system.

processes, water, material use, construction, as well as operations and maintenance. In fact, greenery is only explicitly mentioned in the Soil + Vegetation section. Even within this section, the rating system does not merely emphasise on green coverage but also the quality of greenery and soil, such as (i) using appropriate plants, (ii) conserving healthy soils and appropriate vegetation, (iii) conserving and using native plants, and (iv) conserving and restoring native plant communities.

City Biodiversity Index

This index is focused on the conversation of biodiversity at the city scale. It is used to measure the urban biodiversity of a large area over time. Also known as the Singapore Index on Cities' Biodiversity, the City Biodiversity Index (CBI) was conceived after a series of workshops organised by the Secretariat of the Convention on Biological Diversity (SCBD) in partnership with the Global Partnership on Local and Subnational Action for Biodiversity and the Singapore government from 2009 to 2011 (Chan et al., 2014).

There are two parts to the index (Fig. 10). Firstly, the profile of the city is to be provided. This includes climatic data, population

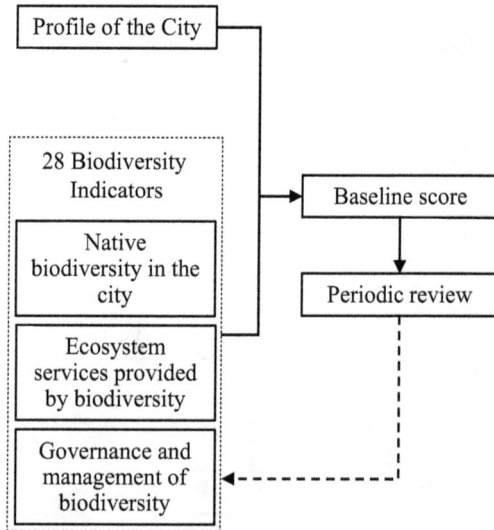

Fig. 10.   Evaluation framework for the City Biodiversity Index.

demographics, biodiversity features, links to biodiversity-related resources, etc. Such information is used to establish the background of the city and to provide reasonable expectations for evaluation.

Next, a total of 28 biodiversity indicators are evaluated. They are grouped into three core components, namely (i) Native biodiversity in the city, (ii) Ecosystem services provided by biodiversity, and (iii) Governance and management of biodiversity. Points are scored by achieving respective targets for each indicator. A total of 112 points can be obtained.

Users of the CBI are encouraged to apply the index to their respective cities, establish a baseline score, and perform periodic re-evaluations to ensure that biodiversity conservation and restoration efforts are materialised. Within each indicator, there are clearly established rationales and quantifiable objectives. The method of calculation and basis of scoring are also provided.

This index does not look solely into green infrastructure but into all major aspects of biodiversity preservation, such as policy and administration. For instance, points are awarded for having a local biodiversity strategy and action plan (Indicator 17) or cross-sectoral and inter-agency collaborations (Indicator 22). In fact, green cover is only featured in a few indicators, such as Indicator 11 (Climate regulation — Benefits of trees and greenery) as percentage tree canopy cover, and Indicator 12 (Recreational services as urban green space per 1,000 persons) (Chan et al., 2014).

Due to the expected large scale of implementation, this index is most suitable for use by governmental bodies with an extended time horizon for evaluation. It is not suitable to be used for building or precinct-scale evaluation.

Neighbourhood Landscape Design Guidelines

The Neighbourhood Landscape Design Guidelines (NLDG) provides a set of guidelines for designing neighbourhood landscapes, with the objective of promoting urban sustainability and liveability by tapping onto relevant ecosystem services. It is neither a checklist nor a scoring system but a conceptual framework derived from key concepts and design principles surrounding the neighbourhood landscape. Design guidelines are provided,

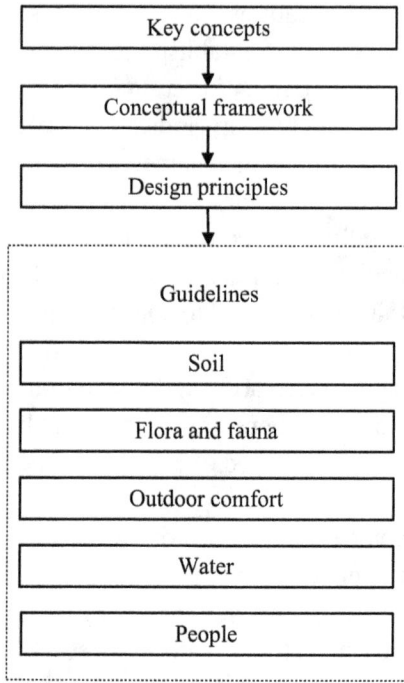

Fig. 11.   Framework for Neighbourhood Landscape Design Guidelines.

together with the rationale and strategies for implementation, targets and performance indicators. These guidelines are grouped into the categories of soil, flora and fauna, outdoor comfort, water and people (Fig. 11). These categories were selected based on their impact and amenity to design (Tan, Liao, Hwang, & Chua, 2018).

Describing landscape structure

For landscape planning and design at the precinct, township or city scale, it is often necessary to quantitatively describe both the composition as well as the configuration of green spaces. These landscape patterns provide some information on the quality of the landscape and its associated ecological attributes (level of biodiversity, cooling potential, etc.).

There is more than one way to characterise landscape structure. This book focuses on the Patch Matrix Model (PMM), as this method can provide a clear delineation of different patches in the surveyed area and employs quantitative techniques that are well-developed and easy to use (Lausch et al., 2015). For the PMM, landscape structure is simplified into three main components: the patch, corridor and matrix. A patch is defined as a homogenous area with distinctive characteristics, such as land cover (e.g., grass patch). Corridors serve to link other patches and are often linear in shape (e.g., stream). They act as conduits to facilitate the movement of flora, fauna and resources. A matrix is the dominant land cover of the surveyed area (e.g., forest). A patch mosaic describes the spatial configuration of a group of patches within a landscape. Landscape metrics are used to quantitatively describe the patch mosaic.

Landscape metrics

Documenting the structure, function and change for a site is essential for understanding its underlying ecological characteristics. Landscape metrics have been widely used in the field of landscape ecology to study landscape function and change.

There are several prerequisites to using landscape metrics. Firstly, geospatial data of the site must be provided. This can be in the form of maps with classified landforms (e.g., dense vegetation, impervious surfaces, water, etc.). A method of obtaining vegetation cover via NDVI calculation has been presented in the preceding section. Subsequently, the map can be processed via programs such as FRAGSTATS "landscapemetrics" package in RStudio to quantify landscape structure (Hesselbarth, Sciaini, With, Wiegand, & Nowosad, 2019). Once the relevant landscape metrics have been generated, they can be used for analysis, such as for longitudinal studies of landscape patterns over time or for evaluation of landscape design schemes on their environmental impact (Uuemaa, Antrop, Roosaare, Marja, & Mander, 2009). The proposed framework for using landscape metrics is shown in Fig. 12.

A video on processing classified images for landscape metrics analysis can be found on www.ugl.sg/book, under the section titled "How to create an NDVI map".

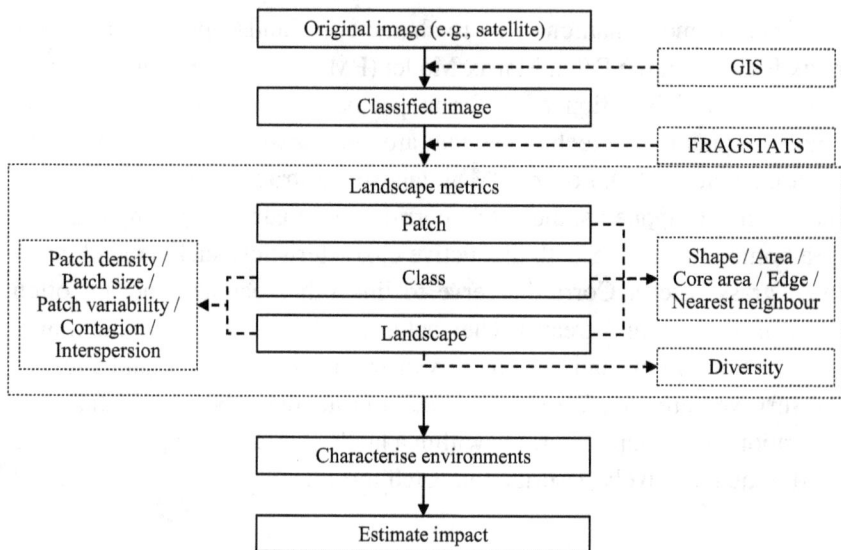

Fig. 12. Proposed framework for using landscape metrics.

Patch, class and landscape

Landscape metrics are grouped into patch, class or landscape levels (Fig. 13).

Patch level metrics describe individual patches in a landscape.

Example: There are 4 patches of B. Their areas are 2, 2, 2, and 9 cells.

Class level metrics describe groups (or classes) of patches with similar characteristics.

Example: There are 4 patches of A. The mean patch size of A is $(1+1+1+9)/4 = 3$ cells.

Landscape level metrics describe the entire area.

Example: There are 10 patches in this landscape. The percentage proportion of C for this landscape is $(9/140) \times 100 = 6.4\%$.

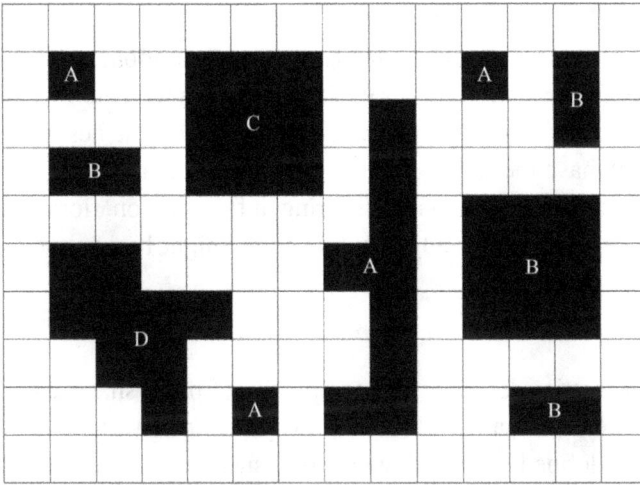

Fig. 13.   Concept of patch, class and landscape.

Commonly used landscape metrics include the following:

Area:
Calculates the area at the patch, class and landscape levels. They provide information on landscape composition, but do not contain details on configuration.

Patch density:
Describes the number of patches per unit area.

Patch size:
Patch size, or specifically, mean patch size, describes the average size of patches over an area. It is a function of the total number of patches in a class over the class area.

Patch variability:
This set of metrics describes heterogeneity in the landscape. For instance, two landscapes may have similar patch densities and mean patch sizes but contrasting levels of variation in patch size. A higher variability would allude to less conformity in pattern, observed at the class or landscape level.

Edge:

This set of metrics describes landscape configuration, in terms of the periphery of the patch. In general, a more convoluted patch would have a higher perimeter (or edge metric). For instance, Patches C and D in Fig. 13 both have the same area, but Patch D has a larger perimeter. Edge effects in landscape can have differing influences on flora and fauna: some species may prefer edges, while some might be adversely affected by them.

Shape:

This set of metrics describes the complexity of patch shape, as a function of its perimeter and area. The complexity of a patch shape is compared with a pre-defined standard geometry, such as a circle or a square (McGarigal, 1995).

Core area:

As opposed to Edge metrics, Core area metrics describe the area within a patch beyond a certain edge distance or buffer, representing both landscape composition as well as configuration. Core areas are sensitive to patch shapes: a convoluted patch may have a large patch area but a very small core area. Exact distance from the patch edge to qualify as a core area can be customised (e.g., in RStudio as input argument).

Nearest neighbour:

This set of metrics describes how far patches are from their nearest neighbouring patch of the same type, based on their distance from edge to edge. They quantify landscape configuration.

Diversity:

Diversity metrics quantify landscape composition and describe the richness and evenness of patch types. Richness refers to the number of patch types found in the surveyed area, and evenness describes the distribution area of the patch types. They are used to estimate diversity via indices such as Shannon's diversity index and Simpson's diversity index.

Note that the indices are used in this context to calculate diversity of the landscape, which hints at the level of biodiversity that may be expected for the surveyed areas. It does not provide information on the exact species of fauna, as well as the nature and extent of their impact (whether they are a desirable species, etc.).

Contagion and interspersion:
This set of metrics describes the extent of the clumping of attributes on a site and the intermixing of classes. It is computed from the frequencies by which different pairs of attributes occur as adjacent pixels on a map.

A detailed list of metrics can be found in the Appendix.

Exercises

Question 1:
What is the Biotope Area Factor (BAF) of this development? Refer to the Ecological Value Factors given below (Fig. 14).

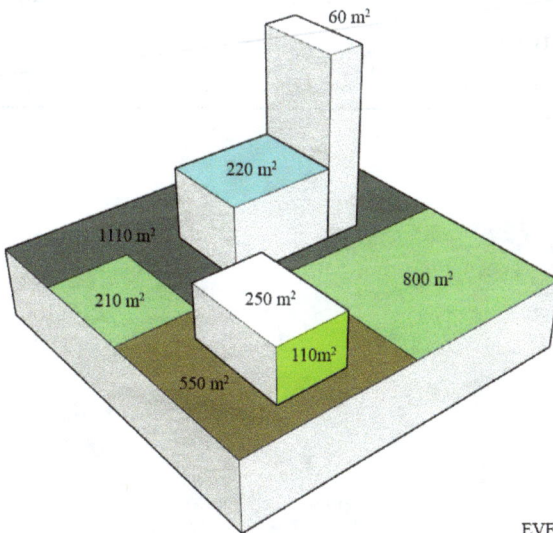

	EVF
☐ = Sealed surface	0.0
■ = Permeable pavement	0.0
■ = Vertical greenery	0.5
■ = Surface with vegetation, connected to soil below	1.0
■ = Semi-open surfaces	0.5
■ = Greenery on rooftop	0.7

Fig. 14.   Hypothetical development for BAF calculation.

Question 2:

What is the Green Plot Ratio (GnPR) for the estate below? Refer to Table 2 and https://www.nparks.gov.sg/florafaunaweb for the Leaf Area Index (LAI) and canopy area values.

Assume site area = 4,500 m²

Category	Subcategory	Quantity
Trees (no.)	Open canopy	6
	Intermediate canopy	12
	Dense canopy	43
Palms (no.)	Solitary	6
	Cluster	0
Shrubs (m²)	Monocot	64
	Dicot	20
Turf (m²)	—	130
Green walls (m²)	—	55

CHAPTER 4

Precinct-scale greenery

This chapter touches on the different forms of greenery and how they are presented at the precinct scale. Urban greenery can exist in many forms and take on a wide variety of combinations. This can make it challenging for designers to decide on how best to implement greenery for their designs. The scale and location of the design are also unique to each project. At the precinct level, it is useful to first categorise greenery into the following:

1. Greenery on ground
2. Greenery with water
3. Greenery on buildings

At this scale, it will be prudent to consider the specific type of ecosystem services that can be acquired from the different forms of greenery. For instance, greenery on ground (e.g., parks) can have a significant influence on air temperature reduction, while greenery with water tends to exhibit more phytoremediative qualities. Similarly, costs associated with the different forms of greenery will also differ: greenery on buildings have specific systems for installation and maintenance regimes that are customised to the building. Finally, key considerations for each form of greenery will be presented to provide a comprehensive framework for evaluation.

Greenery on ground

Greenery on ground is the most common form of greenery to be found in urban areas. Large green spaces exist as forests, parks and gardens, while smaller green spaces take the form of pocket parks, sidewalk planters, roadside hedges, etc. It is also possible to find trees planted in isolation, but a typical landscape arrangement comprises a combination of grass, shrubs and trees.

Implementation framework

Greenery at the ground level can provide a whole host of benefits for the urban environment. However, these benefits may be trait-specific: mean radiant temperature reduction through shade provision is much more tangible for tall trees with large canopies. Therefore, a turfed area will have significantly lesser temperature reduction potential compared to another area of similar size but covered with large trees. The costs of planting and maintaining turfs and trees are also different. Trees have to be inspected and pruned regularly to minimise the danger of falling branches and upheaval, while turf just needs to be mowed. Some key considerations for installing greenery on the ground include plant selection, placement, opportunities and issues arising from the planting site and maintenance strategy (Fig. 15).

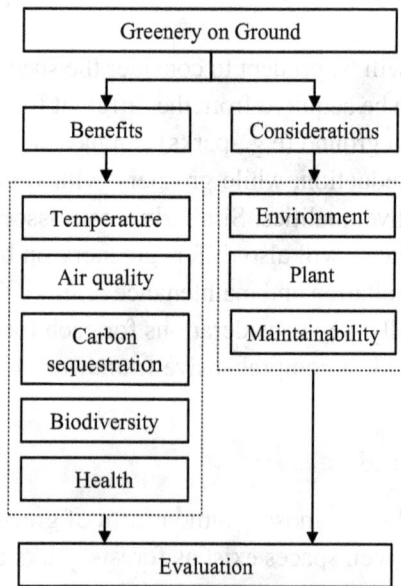

Fig. 15.   Implementation framework for greenery on ground.

Benefits

Temperature

Temperature reduction (or thermal regulation) is a tangible ecosystem service that greenery can provide and is a commonly cited benefit for having greenery in urbanised areas (Wong et al., 2021).

It is important to understand that there are three forms of temperature affected by the presence of greenery (Fig. 16). The first form is mean radiant temperature, which refers to the temperature due to radiation (e.g., direct solar radiation from the sun). Greenery can reduce mean radiant temperature through shade provision. Rays of the sun are blocked by the plant canopy, drastically reducing solar radiation exposure and mean radiant temperature. The second form of temperature is surface temperature. With greenery, less radiation from the sun will reach hardscapes such as roads and pavements. This will lead to less heat gain for these surfaces and a lower resultant surface temperature. In addition, evapotranspiration in

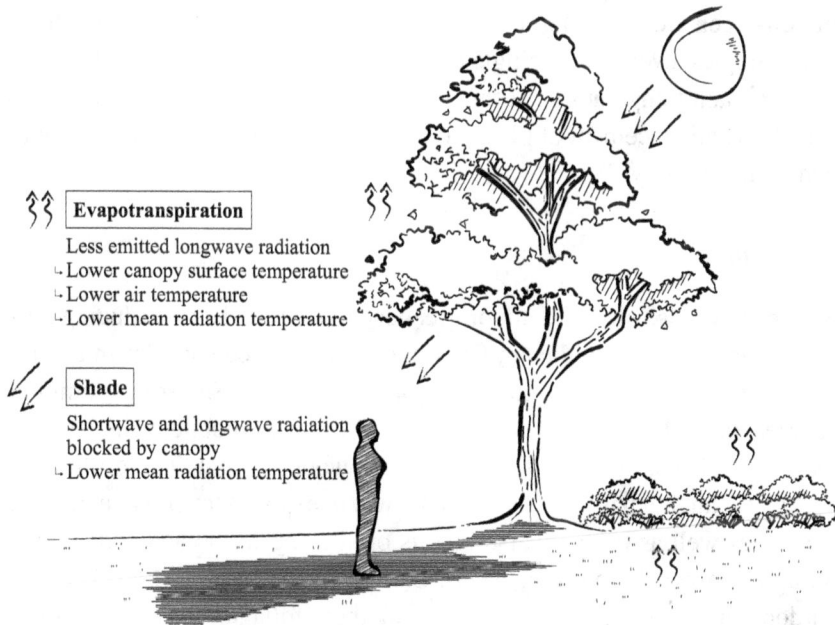

Evapotranspiration
Less emitted longwave radiation
↳ Lower canopy surface temperature
↳ Lower air temperature
↳ Lower mean radiation temperature

Shade
Shortwave and longwave radiation blocked by canopy
↳ Lower mean radiation temperature

Fig. 16.   How greenery can reduce temperature.

plants leads to an increase in latent heat of vaporisation. Leaf and soil surfaces become cooler as a result. The final form of temperature is air temperature. As leaf and soil surfaces become cooler, they reduce the temperature of the surrounding air. The cooler air is circulated via convection to other areas. In this manner, greenery can not only reduce the air temperature within green spaces but also cool down non-green areas in close proximity (Monteiro, Doick, Handley, & Peace, 2016).

Understanding the different forms of temperature and the respective mechanisms for cooling in the presence of greenery are crucial as the mode and magnitude of cooling is highly contextual and dependant on the scale of greenery. For large green spaces such as parks, there are more opportunities to reduce air temperature, both within as well as for areas outside of the green spaces. The high foliage density and substantial canopy cover from trees provide spaces for air to be cooled and carried away via natural ventilation. For small green spaces such as pocket gardens, cooling via air temperature reduction may be less pronounced, as the cooler air may be quickly mixed with the surrounding hot air. When vegetation is placed inappropriately, air temperature can even be higher than in non-green areas (Chang, Li, & Chang, 2007). Nevertheless, shade from tree canopies, even from a single tree, can significantly reduce surface and mean radiant temperature. At this scale, attention should be placed on the selection and placement of plants to provide adequate shade for areas with pedestrian activity.

Air quality

Studies have also shown outdoor greenery to be effective in reducing particulate matter (PM) in the air, PM refers to suspended particles in the air. These particulates can be either in solid or liquid form. Some common sources of PM include dust generated from automobile vehicles and industrial plants, as well as from burning wood or coal. These ultrafine particles can be harmful to humans when exposed over an extended period, as well as have adverse effects on the vegetation and ecosystems (Grantz, Garner, & Johnson, 2003). Greenery can reduce PM levels in the outdoor air by trapping particulates in their foliage. Therefore, larger plants with big, dense canopies should be selected for more effective reduction (Grzędzicka, 2019). A cluster of plants consisting of trees and shrubs with a complex structure and composition would also be more effective than single trees spaced widely apart (Vieira et al., 2018).

Carbon sequestration

Greenery plays a significant role in reducing atmospheric carbon via the photosynthetic process and by storing carbon in their biomass as part of the growth cycle (Nowak & Crane, 2002). Urban greenery (i.e., greenery in the urban environment), on the other hand, may have a lesser impact on sequestration due to their smaller size, plant composition and maintenance regime (Velasco, Roth, Norford, & Molina, 2016). Understandably, the size and density of greenery must be substantial to mitigate anthropogenic sources. Gratani, Varone, and Bonito (2016) found that urban parks can contribute to carbon sequestration and that the rate of sequestration was correlated to the plant's LAI. The highest rate of sequestration was detected in the woods and the lowest at the lawns, suggesting bigger plants with larger biomass to be more effective. Plant management is another crucial factor that may undermine carbon sequestration ability. Over-pruning of tree canopies, for instance, may severely undermine sequestration potential (Gratani & Varone, 2007).

Biodiversity

Rapid urbanisation requires clearing forested areas and can lead to loss of habitat for animals, reducing flora and fauna count. Having greenery in the urban landscape can help to preserve biodiversity. Trees, shrubs and turf help form natural habitats for animals and provide a reliable source of food and shelter. At the same time, flora and fauna diversity promotes resilient green spaces (Gopalan & Radhakrishna, 2022).

The best-case scenario would arguably be for all green spaces to be linked, so that accessibility is maximised for animals. Therefore, when designing green spaces with the intention of promoting biodiversity, it is important to first consider the role of the space at the precinct scale: Is there potential for the green space to function as a linkage between other green spaces? If it is indeed possible, the type of fauna should be identified so that the corresponding flora can be recommended. For instance, when designing habitats conducive to colugos (flying squirrels), tall trees are needed to facilitate movement via gliding from trees. Butterfly and bee-attracting plant species should be considered if the intention is to increase the said population.

The maintenance strategy of green spaces can also influence the level of biodiversity in a given area. Some studies have shown less-manicured green spaces to be more effective at promoting biodiversity than in those that are regularly pruned (Hwang, Yue, & Tan, 2017). The lesser need for maintenance would also lead to a smaller carbon footprint and more sustainable practices. However, due to the "wild" and "unmanaged" look of such spaces, some effort is required to promote the acceptance of such spaces in urbanised areas (Hwang & Roscoe, 2017).

Health

Incorporating green spaces in urban areas can have appreciable restorative effects on the overall health of their inhabitants. Large green spaces such as parks serve to promote active mobility for citizens and offer a viable alternative to travelling by car, increasing the chances of engaging in physical exercise, reducing exposure to anthropogenic pollution and chronic-degenerative diseases, making cities more liveable. There is some evidence that greenery has a positive influence over the quality of window views in high-rise neighbourhoods (Olszewska-Guizzo, Escoffier, Chan, & Tan, 2018).

Greenery forms contemplative landscapes that can improve mental health (Olszewska-Guizzo et al., 2022). With physiological observations (e.g., brain activity), some studies have found that different landscape compositions can have differing impacts on people (Olszewska-Guizzo, Sia, Fogel, & Ho, 2020). In this regard, it is useful to consider an assessment of landscape design proposals using metrics such as the Green View Index (GVI), where a series of photographs taken at eye level are used to estimate green coverage at any given point (Yang, Zhao, Mcbride, & Gong, 2009). This can help make landscape design for improving mental health more objective.

The COVID-19 pandemic has further demonstrated the urgent need for adequate green space provision in highly built-up environments (Olszewska-Guizzo et al., 2021). During periods of lockdown, restrictions to movement can lead to stress and drastic changes in people's moods. Self-reported surveys conducted in one study noted that access to green spaces and natural elements were perceived to contribute to urban well-being during the pandemic (Shentova, De Vries, & Verboom, 2022).

Considerations

Environment

Climate

Climate refers to ambient conditions where planting occurs. This is the key determinant for plant survival. Before deciding on the type of plants for your design, it is important to have a clear understanding of prevailing environmental conditions, as it will influence the choice of plants suitable for your site. If ambient conditions are not ideal for growth, even resilient plants will experience physiological damage. This may result in excessive maintenance or frequent replacement, eventually leading to removal or redesign. Climatic information should consist of the following:

1. Solar irradiance. The main driver for photosynthetic activity. Lighting levels at the ground level tend to be in abundance and do not fluctuate as drastically as greenery on buildings. This may change as green spaces get closer to tall structures or trees with large canopies are introduced.

2. Rainfall and wind. Influences plant selection in terms of their tolerance to drought conditions, high wind speed, as well as irrigation regime of the turf, shrubs and trees. Seasonal changes (e.g., monsoon) with higher magnitudes of wind speed and rainfall should also be noted.

3. Air temperature and relative humidity. Limits the choice of plants that are adaptable to ambient conditions. Data for the entire year should be used, to account for seasonal variations (summer to winter).

Monthly climatic data can be obtained via weather websites for free. This level of resolution is sufficient for most ground-level landscape design projects. A higher resolution (e.g., daily or hourly data) may be required for other forms of landscaping projects, such as skyrise greenery, where microclimatic conditions vary significantly according to factors such as building orientation, self-shading, overshadowing, etc.

Space

The next limiting factor for having greenery on the ground is space. For plants to thrive, sufficient space must be provided both above and below ground. Spatial allocation for plants above ground includes provision for both height and width. Care should be given to avoid possible clashes with overhead wires, lamp posts or signages as trees grow taller over time. Since tree canopy can increase quite drastically as well, designers have to ensure that overgrown canopies will not affect buildings or building systems in close proximity. For roadside trees, the canopy must not grow too large to obstruct visual access for motorists. There must also be sufficient space and structural support in the ground for plants to grow well. The planting area should preferably be wide and deep enough for tree roots to achieve proper growth and anchorage. In natural areas, there is less restriction for root propagation. However, in the urban environment, planted areas can be allocated and carved out of existing urban terrain (e.g., sidewalk). This creates some restriction to the size of trees that can accommodate such planted areas. For cases where there is insufficient soil volume on the ground to support tree growth, raised-bed planting may be a viable alternative. Sufficiently large containers are also a convenient solution to having trees and provide flexibility for relocation if necessary (Fig. 17).

Trees require sufficient time to grow and for ecosystem services to be realised. For instance, significant temperature reduction via shade provision from the tree canopy is only possible when the tree has matured. Due to urban redevelopment (e.g., road widening), mature trees may have to be cut down as they cannot be relocated without sustaining severe root damage. For such cases, containerised trees planted in the ground can help to protect root balls and allow for ease of movement at a later date (NParks, 2011).

Fig. 17.   Trees planted in large containers.

Soil

Besides having a conducive climate and sufficient space, access to good soil is another key consideration to growing plants well. There must be sufficient soil to provide proper anchorage for plants. Nutrients required for plant growth are absorbed from the soil through the roots and into other parts of the plant. Therefore, both quantity as well as quality of soil have to be good enough to support plant growth.

Soil mass is a three-phase system consisting of solid, liquid and gaseous matter (Punmia & Jain, 2005). The solid portion of soil comprises a mixture of sand, silt and clay. They are the same minerals but differ in size, with sand particles being the largest (1 mm to 2 mm in diameter), followed by silt (0.05 mm to 0.02 mm) and clay (<0.02 mm) (Trowbridge & Bassuk, 2004). The composition of these three minerals (sand, silt and clay) in a soil sample determines its texture and denomination (Fig. 18).

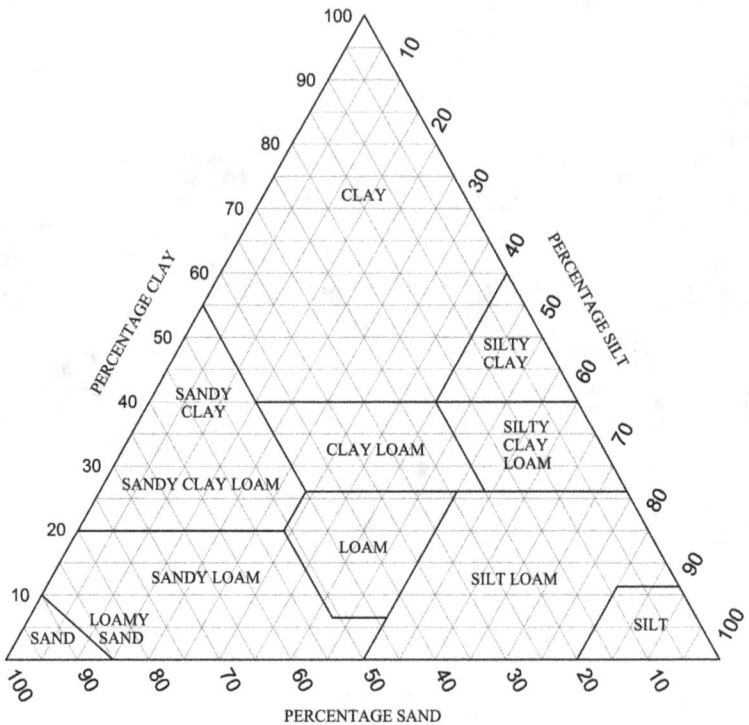

Fig. 18. Soil triangle. Adapted from (USDA, 1993).

Some factors that can influence soil quality include:

1. Texture

Soil texture provides an indication of its composition, structure, density and physical properties, such as water retention and drainage capabilities. Sand is much coarser than clay and is much weaker at water retention. A balance between aeration, water drainage and retention must be struck for plants to grow well. Problems such as leaching may occur when the soil is too porous. Soil texture can be determined both qualitatively by checking the texture of a soil sample by touching it and examining its texture (Thien, 1979) or quantitatively via a percolation test, where a hole is dug, and water is poured into it until the saturation point. The height of water is measured over time to estimate the drainage capability (Peterson, 1980).

2. Density

Soil density is defined by its mass over volume. Soil density is important as it determines the extent of root penetration through the soil. If the soil is too dense, there may be little to no space for roots to grow. This may severely impact plant growth, as water and nutrient uptake will be limited (Soane & van Ouwerkerk, 2013). Soil density can be measured with a penetrometer, which measures the force per unit area required to insert a probe in the soil (Sanglerat, 2012). In the urban environment, soil density may change over time due to compaction from foot traffic, loading from heavy machinery or vehicles, etc.

3. Chemical composition

The main source of mineral nutrients from plants is from the soil. Nutrients are absorbed in soluble, ionic forms and uptake is dictated by soil acidity or alkalinity (pH) (Baligar, Fageria, & He, 2001). Soil pH in urban areas may be contaminated by erosion of building materials such as limestone or concrete through weathering. Soil used for planting may be mixed with non-soil material such as construction debris. This may lead to changes in chemical composition and nutrient uptake in plants (Jim, 1998).

Due to challenges in the urban environment, there may be issues with soil quality for planting. In such cases, the following interventions may be considered:

1. Structural soil

Chances of soil compaction can be reduced by using structural soil, which consists of a combination of stone and soil mixed to form a load-bearing stone lattice, with voids that are interconnected and partially filled with soil to facilitate water, air and root movement (Fig. 19). When properly mixed and with the right ratio, the soil will be subjected to less compaction over time and promote better root growth (Grabosky, Bassuk, & Trowbridge, 2002). There are many variations of structural soil mixes, and their performances may vary. The recommended benchmark would be CU-Structural Soil, developed by Cornell University's Urban Horticulture Institute in the 1990s, which has undergone extensive research and countless tree-planting installations (Bassuk, 2008).

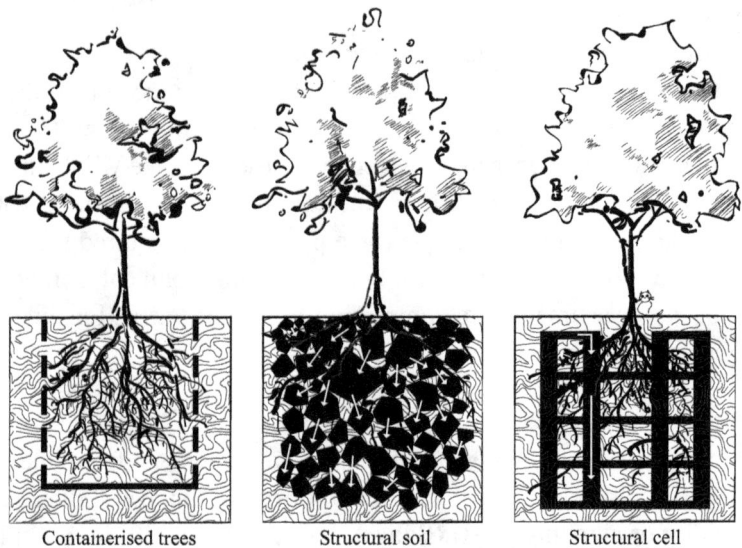

Fig. 19.   Containerised trees, structural soil and structural cells, with arrows showing load transfer.

2. Structural cell

Structural cells consist of modular blocks designed to take pavement loading and voids to hold uncompacted soil as well as manage stormwater runoff (Fig. 19). Since the cells are modular, they can be customised accordingly to accommodate site conditions. Deeper root penetration via the uncompacted soil will improve anchorage and reduce chances of roots growing towards the surface and causing damage to pavements (Page, Winston, & Hunt, 2014). In a study spanning six years, Ow and Ghosh (2017) noted a significant increase in trunk diameter and height of *Samanea saman* and *Peltophorum pterocarpum* species planted in structural cells compared to plants in normal soil conditions.

Plants

Plant selection should commence only after assessing relevant climatic, spatial constraints and soil conditions. Selection, as well as placement, should be conducted in view of plant requirements and ecosystem services.

There must be adequate lighting, water and nutrients for plants to survive and thrive. These resources are abundant in nature but may be in short supply in the urban environment, due to influence from the built morphology and impervious surfaces. Hence, plants should be selected to match the prevailing climate, spatial and soil conditions. Some common classification of plant requirements can be found in Table 4.

Table 4. Common classifications of plant requirements (Boo, Omar-Hor, Ou-Yang, & Ng, 2003).

Light	Water	Special care
Full shade	Little water	Occasional misting
Semi-shade	Lots of water	Prefers cool environment
Full sun	Moderate water	—

After addressing plant requirements for survival, attention should be paid to the ecosystem services they can provide. For instance, if temperature reduction via shade provision is the desired ecosystem service, large canopy trees should be specified. If aesthetics take priority, flowering plants and brightly coloured foliage can be considered.

Lastly, potential disservices should be avoided: Plants that may trap water and encourage mosquito breeding (e.g., *Bromeliads*) should be avoided. Similarly, tree and shrub placement that blocks wind and reduces outdoor ventilation may negate intended ecosystem services such as temperature reduction and needs to be re-evaluated.

Maintainability

Maintenance for green spaces include trimming of trees, shrubs and turf, removal of debris, application of mulch and other chemical treatment.

A sound maintenance or management strategy for ground-level green spaces is important for a multitude of reasons. Firstly, an efficient maintenance plan optimises its use of resources such as fuel and labour, saving costs for the operator. Secondly, it will reduce risks and improve onsite safety levels for green space users (e.g., incidents of users getting hurt by falling branches or debris will lessen). Thirdly, it will ensure that the intended ecosystem services specified in the design stage can be realised to their full potential. For instance, tree canopies that are over-pruned will not have enough canopy cover to ensure sufficient shade provision (Fig. 20). Finally, it may lessen the chance of incurring unintended

Fig. 20. Over-pruned trees resulting in a lack of shade provision. Climate regulation as an ecosystems service is nullified as a result.

disservices, such as increasing the carbon footprint instead of reducing it through an inappropriate maintenance strategy (Velasco et al., 2016). Landscape maintenance operations are influenced by environmental, political, social and financial factors (David, Michelle, Neil, & Kevin, 2016). Sustainable management of UGSs requires a coordinated effort between multiple stakeholders at the design, construction and maintenance phases. Agendas of the project owners, designers, consultants, contractors and maintenance teams should be clearly communicated and be in alignment for sustainable management to be possible (VanDerZanden & Cook, 2010). A comprehensive set of sustainable landscape management strategy for various forms of urban greenery such as lawns, parks, roadside plant, waterway planting and spontaneous vegetation can be found in CUGE (2015).

Greenery with water

Part of the urbanisation process involved clearing natural areas such as forests and replacing them with impervious surfaces such as concrete pavements and tarmac roads. This severely diminishes the porosity and water retention capability of these areas. As a result, highly urbanised areas can be prone to flooding if the conveyance and discharge of stormwater occur at a slower rate than incoming rainfall. Flooding in highly urbanised cities such as Jakarta and Bangkok has become increasingly common over the years in light of extreme weather events caused by climate change, resulting in considerable damage to the property and livelihood of city dwellers.

One method of mitigating the risk of urban flooding is to adopt a Water Sensitive Urban Design (WSUD) approach, which incorporates planning and design principles to address the issue of stormwater in built environments. It employs a design methodology in which water is retained at its source, and the speed of its distribution is slowed down. This includes adaptive approaches when water accumulates at the sink. It is markedly different from the conventional strategy of channelling runoff through pipes and drains into centralised catchments such as reservoirs or constructing flood walls and dams as flood mitigation measures.

Very often, these strategies employ a naturalised approach, with greenery combined with water elements (i.e., blue-green infrastructure). The Sponge City concept, for instance, shows how urban planning can be reimagined with blue-green features to improve the hydrological aspects of urban environments (Chan et al., 2018).

In addition to flood alleviation, several other ecosystem service benefits, such as temperature regulation, water harvesting and purification, can be tapped on. Numerous biodiversity and social benefits have been documented for blue-green infrastructure (Brears, 2018). In light of a more sustainable and naturalistic approach to designing urban areas, it is becoming a trend to incorporate water features in a more naturalistic manner, with greenery playing a major role both in the aesthetics as well as functional aspects of water-based features (Well & Ludwig, 2020). It is rare to find greenery interfacing with water features on buildings, as they can be quite complex systems that require sufficient space to function. Such systems are more commonly found at the ground level, in the outdoors and can be close to buildings. They can exist as standalone features (a swale) or an entire landscaped area (rain garden).

When planning or designing for blue-green infrastructure, their functional contribution should be clarified. Typically, it relates to ecosystem services of flood alleviation, water retention and purification. Different combinations of water and plants can also provide benefits such as food provision, habitat and biodiversity preservation.

In terms of flood alleviation, blue-green infrastructure can exist on buildings as green roofs, sky terraces, planter boxes, and to a lesser extent, green walls. At the ground level, landscaping around buildings can include swales, rain gardens, sedimentation basins, constructed wetlands and cleansing biotopes. Some treatments are designed for the conveyance of stormwater runoff, while others serve to retain water to reduce stress on the downstream stormwater management system.

Using flood alleviation as a starting point, other ecosystem services can be considered. This may involve a careful selection of vegetation to be planted for the blue-green setup. For instance, flowering plants can be planted alongside swales to form ecological corridors to promote biodiversity. Larger plants, such as trees, can be considered for shade provision.

Common treatments suitable for deployment in tropical conditions are discussed in the following section.

Implementation framework

There are many ways of incorporating blue-green infrastructure in urbanised settings. They can exist as standalone features or as one part of a larger, more holistic stormwater management strategy. Successful implementation of blue-green infrastructure requires proper planning, design, engineering and installation of the respective systems. Some useful references can be found in Table 5.

At the planning and design stage, some important points should be considered. The reason for opting to have a blue-green feature must be clear. Besides enhancing the aesthetic qualities of the area through landscaping, designers may consider the potential ecosystem service benefits that may be provided. Flood alleviation is a potential benefit from having blue-green infrastructure. Designers may consider addressing the issue of flooding in many ways, such as by slowing the flow of water or retaining it on site, to reduce the overall strain on the urban drainage system. Other benefits, such as biodiversity preservation or thermal regulation, can also materialise with a complementary plant palette. The proposed framework for implementing blue-green infrastructure can be found in Fig. 21.

Table 5.   References for blue-green infrastructure design.

Topic	Reference
Design guidelines for blue-green infrastructure	Public Utilities Board (PUB), Singapore. *ABC Waters Design Guidelines*. 4th Edition. 2018.
Detailed design specifications for blue-green infrastructure	Public Utilities Board (PUB), Singapore. *Engineering Procedures for ABC Waters Design Features*. 2018
Implementation of blue-green infrastructure in major cities around the world	Brears, Robert C. *Blue and Green Cities: The Role of Blue-Green Infrastructure in Managing Urban Water Resources*. Springer, 2018.

```
┌─────────────────────────────────────────┐
│          Greenery with water             │
│       (Blue-green infrastructure)        │
└─────────────────────────────────────────┘
              │
              ▼
┌──────────────────┐    ┌──────────────────────────┐
│ Flood alleviation│───▶│  Intervention strategies │
└──────────────────┘    └──────────────────────────┘

                    ┌──────────┐   ┌──────────────┐
                   ▶│  Source  │──▶│  Decelerate  │
                    └──────────┘   │  water flow  │
                    ┌──────────┐   │              │
                   ▶│  Pathway │──▶│              │
                    └──────────┘   └──────────────┘
                    ┌──────────┐   ┌──────────────┐
                   ▶│ Receptor │──▶│ Retain water │
                    └──────────┘   └──────────────┘

┌──────────────────┐    ┌──────────────────────────┐
│  Other benefits  │───▶│ Identify ecosystem services│
└──────────────────┘    └──────────────────────────┘

                        ┌──────────────────────────┐
                       ▶│ Air and water purification│
                        └──────────────────────────┘
                        ┌──────────────────────────┐
                       ▶│    Climate regulation    │
                        └──────────────────────────┘
                        ┌──────────────────────────┐
                       ▶│        Well-being        │
                        └──────────────────────────┘
                        ┌──────────────────────────┐
                       ▶│       Biodiversity       │
                        └──────────────────────────┘

┌──────────────────┐    ┌──────────────────────────┐
│  Considerations  │───▶│        Evaluation        │
└──────────────────┘    └──────────────────────────┘
```

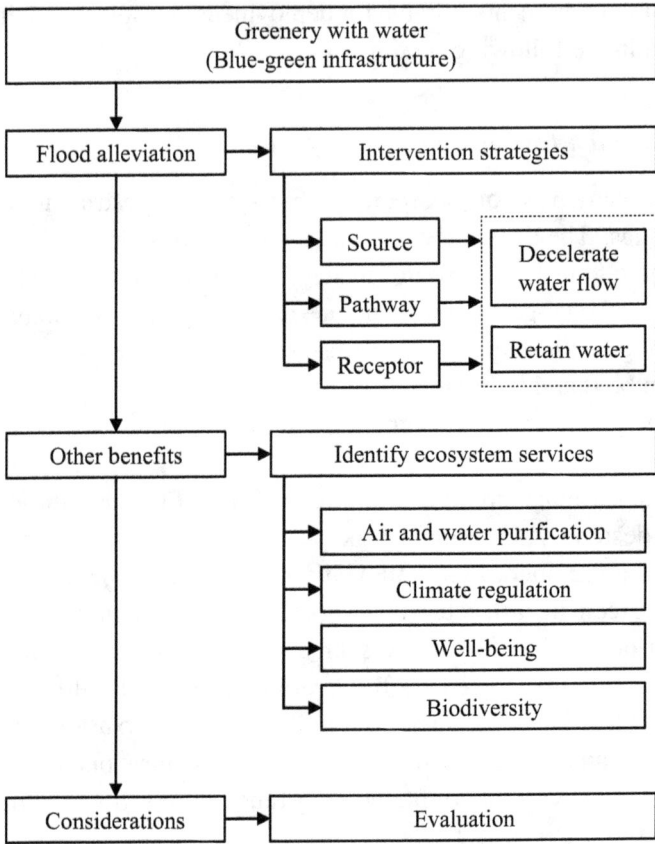

Fig. 21.   Implementation framework for blue-green infrastructure.

Flood alleviation

Some degree of flood alleviation may be achieved by implementing blue-green infrastructure. In addition, expanding the capacity of man-made drains and canals, several intervention strategies involving blue-green infrastructure can be considered. There are many variations to the eventual proposed blue-green solution, but most revolve either around retaining the water at its source or reducing the flow speed. Broadly speaking, the strategies are targeted at three specific points, namely:

1. At the source of rainfall, where rain first comes in contact with the ground or built surface. Stormwater can be stored and slowly released to lessen the strain on drainage systems. Common architectural structures can be used as water retention points at the source. These can be water detention tanks or blue-green infrastructure such as green roofs, sky terraces, planter boxes, green walls and rain gardens.

2. Along the drainage pathway, where stormwater is in conveyance to its intended destination. These are drainage networks designed to provide more drainage capacity than conventional concrete drains or canals and reduce the speed of drainage flow and other ecosystem services. Some examples include vegetated swales and bioretention swales.

3. At the receptor, where stormwater has reached its destination. These can be low-lying areas of water sheds that are more vulnerable to flood risks or intended destinations such as reservoirs. A combination of treatments can be applied at the receptor in order to retain stormwater runoff for as long as possible. Reservoirs and massive water detention tanks, such as the Stamford Detention Tank in Singapore or the Metropolitan Area Outer Underground Discharge Channel in Tokyo, Japan, are prominent infrastructure projects to manage flooding in highly urbanised areas. This is also possible with blue-green infrastructure, such as constructed wetlands and sedimentation basins, where water is retained for an extended period of time.

Other benefits

Water and air purification

Besides flood alleviation, blue-green infrastructure can help to enhance water conservation efforts. As a valuable, life-sustaining natural resource, water harvested by blue-green infrastructure can help abate water scarcity in highly populated urban areas, improving water supply and reducing the need for additional man-made water conveyance, storage and treatment infrastructure. Vegetation along waterways can help intercept runoff and trap pollutants before they are washed away to designated pathways and receptors. Plants and substrates can filter and break down contaminants commonly found in urban runoffs, such as heavy metals. This process occurs naturally, reducing the need for mechanical efforts that require additional resources and costs (Foster, Lowe, & Winkelman, 2011). Vegetation provides protection along waterways through their roots and foliage. This reduces the chances of soil erosion and sedimentation for waterbodies, maintaining flood protection capacity as well as cleansing functionality (Cameron et al., 2012). Greenery along waterways and waterbodies can contribute significantly to the improvement of urban air quality. Firstly, plants can reduce CO_2 and increase O_2 concentration via the photosynthesis process. Plant foliage can trap dust particles and absorb gaseous pollutants (del Carmen Redondo-Bermúdez, Gulenc, Cameron, & Inkson, 2021; Han et al., 2022).

Climate regulation

By having green coverage along waterways, blue-green infrastructure can help improve ambient thermal conditions in the urban environment. This is achieved through the combined effects of shade provision and plant evapotranspiration. To a certain extent, the albedo of the urban environment is higher with greenery (Wong, Tan, Kolokotsa, & Takebayashi, 2021), resulting in less heat trapped in the urban surroundings. Introducing blue-green infrastructure is an excellent way of mitigating the UHI effect. This form of passive intervention, if conducted on a large scale, has been documented to lower cooling energy demand and reduce carbon emissions (Xi, Ding, Wang, Feng, & Cao, 2022).

Well-being

Blue-green infrastructure can promote well-being for urban inhabitants. Serving as multifunctional spaces, they can provide the benefits of flood alleviation, purification and thermal benefits, as well as gathering spaces for social functions. Naturalistic design features (e.g., pleasantly landscaped swales) can encourage patronage of these areas and promote healthy lifestyles. Large-scale blue-green infrastructure often doubles up as parks, giving residents more access to physical and recreational activities. This can translate to higher rates of physical activity, such as walking, jogging and cycling, which improves physical well-being (de Jong, Albin, Skärbäck, Grahn, & Björk, 2012).

The restorative effects of parks and gardens can also help to alleviate fatigue and stress for their users. This may be achieved through a combination of physical/social activities (exercises or having a picnic), as well as access to pleasant views and soundscapes with calming, naturalistic ambience (Bratman et al., 2019). In beautifying the urban area with blue-green infrastructure, better interaction can be fostered between its inhabitants (Lamond & Everett, 2019). This can take the form of team-based activities, or community gardens, where neighbours can work together to maintain their plots. The increased activities can improve social cohesion and boost social well-being for those involved.

Biodiversity

Blue-green infrastructure comprises blue elements such as swales, rivers ponds and wetlands and green elements such as parks, gardens, trees and shrubs. These natural elements support life by providing food and shelter for animals, leading to the preservation of biodiversity in urban environments. With proper plant selection, blue-green infrastructure can function as biodiversity conservation and promotion areas (Krivtsov et al., 2022) or riparian corridors to connect other green areas (Keten, Eroglu, Kaya, & Anderson, 2020).

Considerations

Several planning and design factors have to be taken into consideration when adopting blue-green infrastructure.

By default, a comprehensive drainage system must be planned for a development project, leading to the designated outlet drainage. Planners may consider utilising blue-green infrastructure for portions of water drainage or retention. This may comprise swales and rain gardens.

Some sites may already be endowed with blue-green features, such as existing greenery that can function as phytoremediation filters, undulating topography that channels stormwater runoff and wooded areas sitting on soil that facilitates water infiltration. These features can be identified at the planning phase for blue-green infrastructure projects.

During the design phase, management and control of flooding are essential. The design should cater to peak flows. Peak runoff needs to be estimated for sizing the overflow system of the blue-green infrastructure (Check Table 5 for references). Subsequently, benefits such as water filtration and purification may be considered. The appropriate plant selection can be made based on the desired ecosystem benefits as well as a thorough inspection of site conditions to ensure suitability. Proper access for maintenance works and risk assessment should be conducted to ensure safety in the construction as well as operations and maintenance phase.

Intervention strategies

Swales

Swales are drainage channels with a mild slope, formed by slight depressions on the terrain. Functioning like conventional drains, they are used to convey stormwater. The mild slope results in a slower conveyance speed, increasing the likelihood of sedimentation. With the slower speed, there are less chances of erosive flows compared to normal concrete drains. Some common examples include vegetated and bioretention swales. Bioretention swales function the same way but with an added bioretention layer at the base to filter stormwater runoff before further conveyance downstream. Modular bioretention swales can be used to facilitate construction.

There is a wide range of applications for swales, such as along roads and in carparks, residential, commercial and industrial areas and parks. Swales beautify the surrounding landscape while detaining coarse particles in the runoff, as a form of pre-treatment for further downstream treatment.

Regular maintenance is required to ensure that conveyance is not impeded by debris, overgrown/unhealthy plants or erosion over time. This includes routine inspection of the longitudinal as well as cross-sectional profiles of the swales.

Bioretention basins

Bioretention basins are vegetated depressions on the terrain used to detain stormwater runoff. Also referred to as rain gardens, they function in the same manner as bioretention swales but do not channel runoff, only storing it instead. Bioretention basins can take on a multitude of shapes and sizes (from planter boxes to large gardens), making them versatile for deployment in urban areas.

Stormwater is detained and treated in bioretention basins via sedimentation, as well as filtration and phytoremediation. Larger sediments can be removed upstream via swales.

Similar to swales, bioretention basins require regular inspection and maintenance to function properly. Plants in bioretention basins serve the important function of maintaining soil porosity and biological uptake. Care should be taken to ensure that they are free from stress, disease or pests.

Sedimentation basins

Sedimentation basins are ponds that temporarily detain and retain stormwater runoff. As the name suggests, sedimentation basins are designed for large sediments to settle. It is a useful upstream treatment feature that can be integrated with bioretention basins or wetlands. They can also help regulate downstream flows during storms by serving as additional catchment areas. When properly designed and maintained, sedimentation basins can be a great addition to the landscape (e.g., a pond) and improve its aesthetic quality. The stored water closer to the surface of the basin may be used for non-portable purposes such as plant irrigation.

For sedimentation basins to function properly, regular checks on sedimentation levels, as well as periodic removal of accumulated sedimentation are required.

Constructed wetlands

Constructed wetlands mimic the function of natural wetlands to treat stormwater, sewage or industrial wastewater. Contaminants in the water, such as suspended particles and soluble pollutants, are removed as they pass through the vegetated waterbody via a combination of physical, chemical and biological processes (Moshiri, 2020). Constructed wetlands generally consist of an inlet zone to remove coarse sediment, a macrophyte zone that is heavily vegetated over a shallow waterbody for removing soluble pollutants and fine particles and a high-flow bypass channel for excess water to flow around the wetland without damaging the vegetation.

Constructed wetlands can vary significantly in size and provide multiple ecosystem services, such as biodiversity preservation and climate regulation (Hsu et al., 2011; Ruiz-Aviles, Brazel, Davis, & Pijawka, 2020). They can be categorised into (i) Surface flow constructed wetlands, where water slowly flows through a shallow marsh of emergent plants; (ii) Subsurface flow-constructed wetlands, where water flows below the ground surface and movement of water is not visible. As such, there are fewer issues with pests or odours (Halverson, 2004); (iii) Constructed floating wetlands, where emergent plants are grown on a buoyant structure, floating on the water surface. The extensive root system acts as a natural filter, cleaning the water (Pavlineri, Skoulikidis, & Tsihrintzis, 2017).

The flow of water in constructed wetlands can be either horizontal or vertical. Horizontal flow constructed wetlands treat water by letting it flow horizontally across the wetlands, through a media consisting of the plant roots and gravel. Cleansing biotopes are a form of vertical flow constructed wetlands. Water is filtered and cleaned as it percolates down through the bed and towards the bottom of the constructed wetland, before it is being drained or recirculated (Vymazal, 2011).

Greenery on buildings

Urban environments are often characterised by a high population density, compact urban forms and high-rise buildings, leaving little room for large expanses of ground level greenery. It is more common to find smaller fragments of greenery interspersed throughout the urban scape in the form of gardens, streetscapes and landscaping around buildings. In view of the physical restrictions, it has become more common to find greenery above ground. Also termed as skyrise greenery, such forms of greenery can exist as sky terraces, planter boxes, potted plants in balconies, green walls and green roofs. In addition to outdoor and semi-outdoor settings, it is also common to find greenery inside buildings.

Due to recent developments in technology, skyrise greenery has become a viable solution to promoting greenery provision in high dense urban areas. Many cities have adopted policies to increase green cover via skyrise greenery (green walls or green roofs) in addition to traditional ground level greenery. The benefits of having skyrise greenery have also been documented in multiple studies. Some examples include temperature reduction, biodiversity preservation and reduction of stormwater runoff (He et al., 2021; Wooster, Fleck, Torpy, Ramp, & Irga, 2022; Zhang et al., 2018).

However, due to multiple factors such as plant palette, system components, loading, orientation and exposure to the microclimate, the requirements for skyrise greenery systems to be successfully deployed can differ drastically between systems. Interactions between greenery and building systems may also lead to unintended consequences and issues with safety and maintainability. The type of ecosystem services that can be provided is also specific to each form of skyrise greenery.

It is therefore important to have a clear understanding of how different forms of skyrise greenery (i.e., green walls and green roofs) work for deployment to be more effective.

Green walls

Green walls refer to vertical surfaces that have plants attached to or are standing against them, either inside or outside the building (Safikhani, Abdullah, Ossen, & Baharvand, 2014). Traditionally, vertical greenery refers to climber plants growing directly on the building façade. With advances in technology, many innovative systems have been developed to support the growth of plants in a vertical array. Green walls are arguably one of the most innovative urban greenery solutions providing environmental, economic and social benefits while increasing green cover in densely built-up cities with limited planting space. Green walls can be broadly categorised into support and carrier systems (Chiang & Tan, 2009).

Implementation framework

When deciding whether or not to install a green wall, it is important to consider several key points to minimise failure in the life cycle of a green wall. The three main considerations are to:

1. Ascertain the benefits of having the green wall. There may be a myriad of reasons for wanting to have a green wall. It may be for aesthetic reasons, to increase the green density of the development, to reduce temperature transmission, etc. After a desired benefit is identified, it may be possible for it to be optimised. For instance, if temperature reduction is the desired benefit, plant selection can be geared towards species of larger canopies or higher transpiration rates for added cooling. Benefits can be classified as direct or indirect: temperature reduction is a direct effect that can be readily observed, whereas energy savings is an indirect consequence of reduced heat transmission into the building, lowering cooling energy.

2. Tabulate the costs of having the green wall. Besides the initial cost of installing the green wall, maintenance costs will aggregate to a significant amount in the lifespan of the green wall. The cost of both installation and maintenance of different green wall systems may differ tremendously. Knowledge in the life cycle cost of the green wall system makes it possible to reduce costs and resources in certain areas via pre-emptive measures, such as designing for ease of maintenance and less reliance on expensive equipment such as boom lifts and cherry pickers, etc. Similar to benefits, costs can be direct (installation costs) or indirect (replacement of plants over time, influenced by design or maintenance regime).

3. Consider factors that will affect green wall installation and upkeep. This section examines factors that will influence processes such as landscape design, plant palette, systems selection and maintenance regime. They are broken down into environmental, system and plant considerations, any of which can influence the overall viability of having a green wall in the long term. By considering these factors and their impact on relevant benefits and costs, the value of the green wall can be optimised.

The proposed framework for assessing the suitability of installing a green wall is shown in Fig. 22. It is recommended that the main criteria of Benefits, Costs and Considerations be evaluated in a comprehensive manner before commissioning any green wall project. Details of each criteria are elaborated in this chapter.

Benefits

Many studies around the world have provided evidence of the multiple benefits that green walls can bring. Where appropriate, these benefits can be used as part of the assessment framework shown in Fig. 22. They can

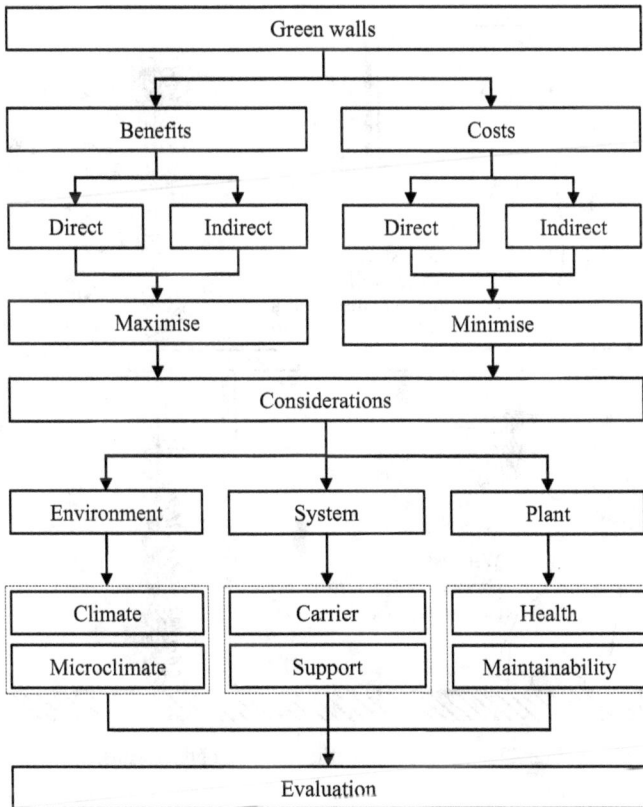

Fig. 22.   Implementation framework for green walls.

also be used to explain to stakeholders why it is beneficial to install a green wall. Some examples are provided below (Fig. 23):

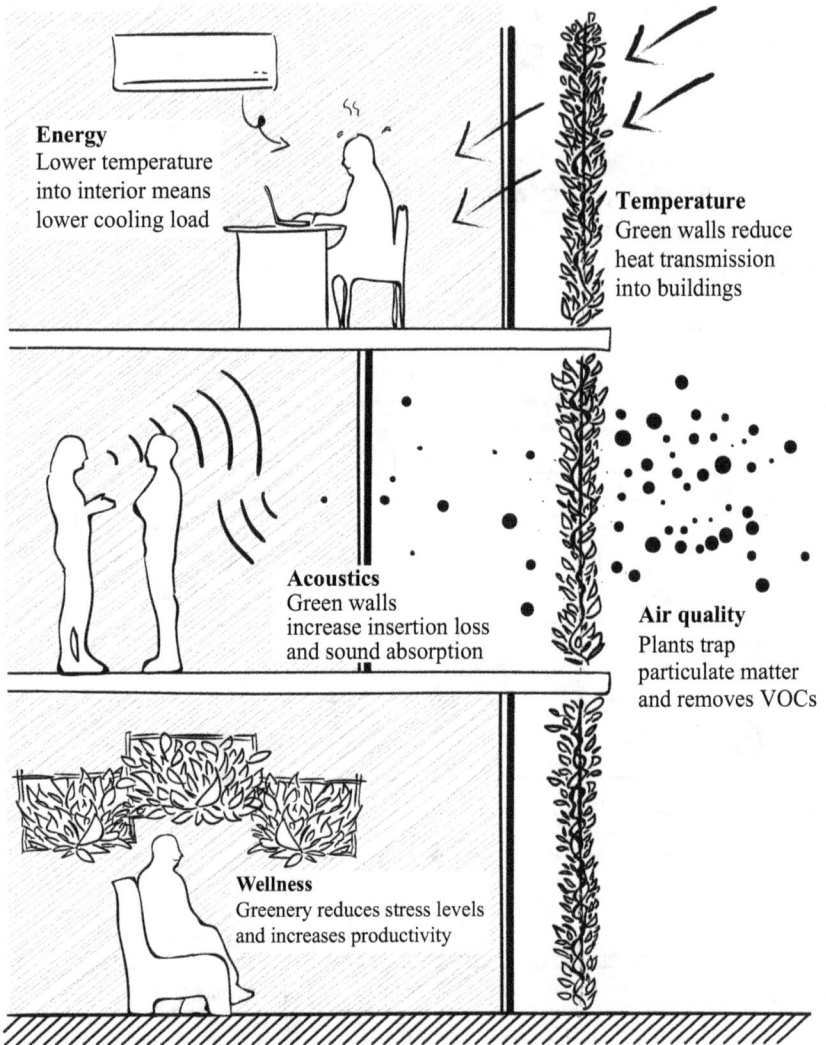

Energy
Lower temperature into interior means lower cooling load

Temperature
Green walls reduce heat transmission into buildings

Acoustics
Green walls increase insertion loss and sound absorption

Air quality
Plants trap particulate matter and removes VOCs

Wellness
Greenery reduces stress levels and increases productivity

Fig. 23.   Potential benefits of having green walls.

Temperature

One major benefit of having green walls is that it can help reduce temperature. As mentioned in preceding chapters, different forms of greenery reduce temperature in different ways. When assessing temperature reduction, it is important to first define the type of temperature in question. Generally, the more common types used for assessment are air, surface and mean radiant temperature. Due to the limited area of green walls, air temperature reduction may only be significant within the foliage and close to the wall. Similarly, mean radiant temperature may not decrease much, as plants on green walls tend to have small canopies and do not provide much shade. Surface temperature, on the other hand, can be significantly reduced as the entire green wall system (plant, substrate and substructure) provides a thick cover over the wall and shields it from direct solar radiation (Bass, Liu, & Baskaran, 2003). Measurements of walls covered with greenery as well as exposed walls have shown a decrease of up to 18.8°C in surface temperature for the former (Wong et al., 2021). In other words, green walls provide an additional layer of thermal insulation for building surfaces.

Energy

When the external wall surface temperature is reduced, less heat will be transmitted into the building. This makes interior spaces cooler and closer to the HVAC set point temperature. As there is less heat to remove, the energy required to cool the interior space will be reduced (Wong, Tan, Tan, & Wong, 2009). In this manner, the presence of green walls can also help reduce energy consumption for the building (Fig. 23). Studies have shown savings of up to 23% when green walls are added onto building façades (Coma et al., 2017).

Acoustics

Green walls can influence acoustic quality by increasing insertion loss and sound absorption. Insertion loss is defined as the difference in Sound Pressure Level (SPL) measured at the same spot, before and after an object is inserted between it and a sound source. Sound absorption is defined as the ability of a material to absorb sound, with "0" being total reflection and "1" being total absorption. One study has shown that for frequencies between 125 Hz and 1,250 Hz, insertion loss can increase by up to 9.9 decibels (dB) in the presence of green walls. In the same study, a green wall was found to have a sound absorption coefficient of close to

0.5, increasing at higher frequencies (Wong, Tan, Tan, Chiang, & Wong, 2010). In the outdoor environment, green walls can potentially reduce urban noises such as traffic noise (Azkorra et al., 2015; Medl, Stangl, & Florineth, 2017; Van Renterghem, Botteldooren, & Verheyen, 2012). The combination of diffracting, reflecting and absorbing sound helps to reduce noise levels and, in turn, improves well-being (Medl et al., 2017).

Air quality

Plants can improve air quality by taking in harmful chemicals in the atmosphere and sequestering them in their roots and cells. Some of these chemicals are broken down by fungi in the substrate, while others may be stored in the plant (Ottelé, van Bohemen, & Fraaij, 2010; Perini, Ottelé, Giulini, Magliocco, & Roccotiello, 2017; Sternberg, Viles, Cathersides, & Edwards, 2010).

Similarly, numerous studies have shown green walls to be able to remove pollutants such as particulate matter (PM) (Pettit, Irga, Abdo, & Torpy, 2017), inorganic gaseous compounds such as nitrogen oxides (NO_x), carbon dioxide (CO_2), carbon monoxide (CO) and sulphur dioxide (SO_2) (Soreanu, Dixon, & Darlington, 2013), as well as volatile organic compounds (VOCs) such as formaldehyde and benzene (Darlington, Chan, Malloch, Pilger, & Dixon, 2000).

Wellness

The high aesthetic value of green walls can improve the quality of human experience in urban environments. Studies have shown greenery to help increase tolerance to pain (Lohr & Pearson-Mims, 2000), reduce stress levels and increase productivity (Lohr, Pearson-Mims, & Goodwin, 1995; Yeom, Kim, & Hong, 2021). Such benefits can help increase the value of the estate. One study noted a 2–5% increase in property value due to the presence of green walls (Perini & Rosasco, 2013).

Costs

One major impedance towards the widespread deployment of green walls, besides spatial limitations, is cost.

Expenditure such as manpower, resources and equipment for installing and maintaining green walls may prove to be prohibitive, even when weighed against potential benefits. While the benefits associated with green walls have been presented in the preceding section, the underlying cost of green walls should also be addressed.

The total cost of a green wall can be broken down into pre-installation, installation, maintenance and disposal costs.

Pre-installation and installation

Pre-installation costs refer to costs associated with the green wall before the construction of the actual green wall. This may include design costs (e.g., for a feature green wall meant to be a showpiece in the lobby) or rectification works such as installing water or electrical points for irrigation pumps and growth lights. For green walls that are meant to be retrofitted onto cladded walls, the claddings may need to be removed prior to the installation of the green wall substructures. Understanding the context in which the green wall is to be installed can help minimise additional costs due to pre-installation works. For instance:

1. Placing a green wall near an existing water and electrical source reduces the need for additional piping and wiring works.
2. Siting a green wall in a space that has access to natural daylight (e.g., from a large window or skylight) reduces the need for growth lights.
3. Selecting a support system instead of a carrier system will result in much less work and loading on the existing wall.

The installation of green walls takes into consideration items such as the initial capital expenditure of the system, including material, manpower and transportation. In this instance, installation costs refer to costs associated with:

1. Building the substructure supporting the green wall system.
2. Installing the green wall system onto the substructure.
3. Installing auxiliary components of the green wall system, such as irrigation system (pumps, pipes and emitters), fertigation system, growth

light systems, etc. These components may be outsourced to third-party vendors.

4. Supply and installation of plants on the green wall.
5. Obtaining approval from authorities (such as PE certification for loading, etc.).

The indicative costs of green walls are shown in Table 6. Over time, the price will tend to decrease. It can be observed that prices differ substantially between systems. This is mainly due to the different green wall systems used, with details covered in subsequent sections.

Table 6. Indicative installation capital cost of green walls (Ho, 2020; Manso, Teotónio, Silva, & Cruz, 2021).

Green wall system	Cost (SGD) / m²	Average cost (EU) / m²
Support	550–800	190
Carrier	850–1,100	848

Maintenance and disposal

In considering the overall Life Cycle Cost (LCC) of a green wall, the largest portion of costs goes to maintenance.

In a study conducted in Singapore, the annualised LCCs of 40 green walls were tabulated (Huang, Tan, Lu, & Wong, 2021). Results showed that maintenance accounts for a major portion of LLC: 72% for carrier systems and 69% for support systems. Initialisation costs consist of items such as green wall structures, irrigation, substrate and plants. This includes the cost of pre-growing the plants in the nursery, labour, fertiliser, irrigation, storage, etc. Installation costs consist mainly of labour and equipment. These costs constitute only a small portion of the overall LCC (Fig. 24).

The high maintenance cost can be attributed partly to expenditure on equipment. For instance, large green walls with a height of more than 3 m may require specialised equipment such as boom lifts, gondolas or cherry pickers for maintenance work such as pruning and plant replacement.

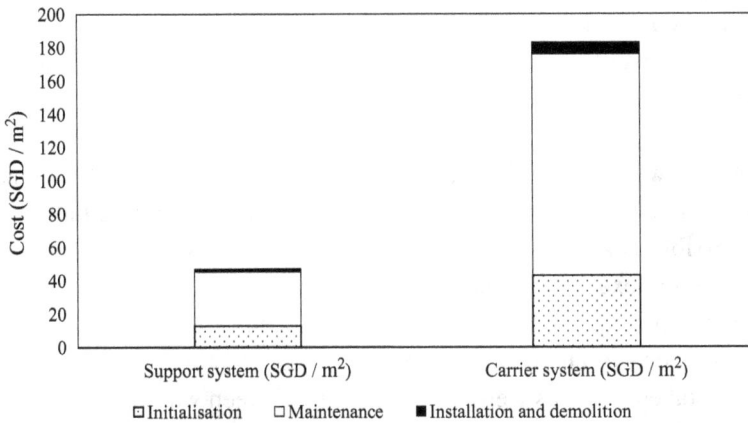

Cost	Support system (SGD / m²)	Carrier system (SGD / m²)
Initialisation	12.92	42.96
Maintenance	32.53	132.89
Installation and demolition	1.70	6.95
Total	47.15	182.79

Fig. 24. Annualised LCC of support and carrier system green walls. (Adapted from (Huang, Tan, Lu, & Wong, 2021).)

Indoor green walls that require growth lights may also require periodical servicing and replacement.

Another significant contributor to maintenance costs is plant replacement, which may occur as often as once per month. In addition to new plants, costs for labour, fertiliser and pesticides also add to the overall plant replacement costs.

Such high costs can be avoided with sensible design choices and informed plant selection to minimise maintenance by adopting the green wall implementation framework shown in Fig. 22.

Considerations

There are three main points to consider when designing for a green wall. They can be broadly categorised into (i) environmental, (ii) system, and

(iii) plant considerations. These points are independent of each other but can interact over time to alter the quality of green walls. They are of equal importance and should be accorded similar weightage during the assessment.

As shown in Fig. 22, the approach adopted in the implementation framework is to maximise the benefits while at the same time minimising any possible costs (or disservices) associated with green walls. This is achieved by examining the key considerations in a holistic manner, including all stages of the life cycle of the green walls.

These points are based on the reviewed literature, interviews with various stakeholders as well as assessments of green walls conducted over a substantial period by the author.

Environment

1. Climate
Climate refers to long-term ambient conditions where the green wall is sited. This is the key determinant for plant survivability. For the purposes of assessment, the term "Climate" here refers to prevailing environmental conditions of the region to which the green wall is situated and should include data on the following parameters:

a. Solar irradiance. The main driver for photosynthetic activity. Through solar irradiance, Photosynthetically Active Radiation (PAR) and Daily Light Integral (DLI) values can be derived.

b. Rainfall. Influences plant selection in terms of their drought tolerance as well as irrigation regime for the green wall.

c. Air temperature and relative humidity. Provides a good indication of the species of plants that can grow well under ambient conditions, without risk of cellular damage due to extreme heat or cold.

When describing the climate of a location, the recommended temporal scale should be in years (not months or weeks). This is to account for the effect of seasonal changes. In tropical regions, there may not be drastic changes in ambient conditions (such as between summer and winter), but there may still be events that significantly affect plants, such as monsoon seasons where periods of heavy rainfall are expected.

There are many ways of acquiring climatic data. Most weather websites provide some form of yearly average climatic information, mainly for people to get a quick sense of the conditions of the place of interest (e.g., www.weather.com). Such information can normally be obtained free of charge and may be sufficient for a cursory understanding of the site. Data of higher resolution, such as daily or even hourly meas-urements, can be purchased through meteorological organisations.

Climate data is normally obtained from weather stations, which is used to approximate the conditions of a larger area (a city or a small coun-try). Therefore, key considerations for assessing climate data are to ensure that the location of the weather station is close to the site of interest and that the resolution of data is sufficient to provide a clear understand-ing of the climate to facilitate design decision making. This may include anything from temporal (hourly or daily data) to spatial to sensor data resolution (accuracy of 0.1°C or 0.01°C).

2. Microclimate

Once climate data is available, designers can begin to filter the list of plants suitable for green walls according to prevailing climatic conditions. Some general exclusions can be expected (e.g., leafy plants cannot survive in desert environments), but there are still many plants to choose from. To further refine the selection of suitable plants in the design palette, it will be necessary to consider the microclimate of the green wall setup.

For the purpose of this assessment, the term "microclimate" refers to the immediate spaces that surround the green wall site. Practically, it should be representative of conditions from 0 m to 50 m away from the site of interest. This is in contrast to climate data, which tends to provide an approximation for a larger region of interest (for a few kilometres). Unlike the prevailing climate of a region, the urban microclimate is het-erogeneous and highly susceptible to influence by its surrounding con-stituents. For instance, there may be high solar exposure for the city in general but low exposure for the green wall site due to self-shading or overshadowing from adjacent buildings. Similarly, an area may have an average air temperature range of 28°C to 32°C but a microclimatic range of 30°C to 36°C due to close proximity to anthropogenic heat sources such as cars or AC exhaust. Microclimatic measurement is not restricted to the outdoor environment: it can be indoors or outdoors, depending on the location of the green wall.

Microclimatic data for green walls typically consists of the following parameters:

a. Solar irradiance. Measurements of solar irradiance at the microclimatic level can help identify site-specific phenomena such as self-shading and overshadowing. In the indoor environment, it can help to quantify daylighting for use in designing supplementary lighting. This can help to reduce energy consumption from growth lights (Fig. 25).

b. Rainfall. Exposure to rain may be affected by building components such as shelters, building convolution and orientation. Effects may vary from excessive rainfall (due to channelling of pipes or shelters) to too little rainfall (blocked by shelters, being situated indoors), all of which will have to be taken into consideration during the irrigation design.

Fig. 25.   Green wall with access to daylight for parts of the day, making artificial lighting less necessary.

c. Air temperature and relative humidity. Provides a good indication of ambient conditions, especially context-specific data such as thermal impact of building materials and anthropogenic heat sources.

d. Surface temperature. Certain façade materials, such as metal claddings being good conductors of heat, can become extremely hot when exposed to direct solar radiation. This may affect support system green walls, in which climber plants come into direct contact with the surfaces, resulting in cellular damage to plants and affecting growth.

e. Wind. Wind profiles can fluctuate wildly in highly urbanised environments. Differences in height, roughness and spatial composition can lead to sudden and drastic alterations in wind speed (e.g., Venturi effect), resulting in sudden gusts of strong winds. Measurements of wind speed around the vicinity can help mitigate risks associated with adverse wind conditions. Several factors can be considered, such as choosing the appropriate green wall system and suitable plants for the wall, as well as whether additional reinforcements to the green wall are needed for ensuring safety (e.g., if the site experiences a typhoon).

Microclimatic data is more challenging to obtain due to site specificity, especially when the building itself has yet to be constructed. Methods of data acquisition are covered in subsequent chapters of this book.

System

1. Carrier system green wall
There are two main types of green wall systems: carrier and support. The design, construction as well as maintenance of these two systems are remarkably different, and selection should be conducted in a holistic manner, as laid out in Fig. 22.

Carrier system green walls are characterised by an even distribution of substrate across the entirety of the green wall. The system is designed to support the weight of the substrate while housing plants securely in carriers. The carriers are modular and attached to a frame anchored onto an existing wall. The modules contain soil or artificial growing mediums such as perlite, foam, mineral wool and felt and may require fertigation systems to balance nutrients and water (Perini, Ottelé, Haas, & Raiteri, 2011). The main components of carrier green walls are as follows (Fig. 26):

a. Substructure. This layer is in direct contact with the existing wall. It has to be securely fastened to the wall to ensure stability and proper load transfer from the plants and substrate to the wall. Typically, metal channels (C-channels) are bolted onto the wall to form a frame. Subsequently, a metal mesh is fastened onto the frame and will be used for the carriers to be mounted on. In some cases, a layer of protection (e.g., water proofing layer) is added between the substructural and existing wall if there are concerns with irrigation leakage or plant root penetration. The eventual

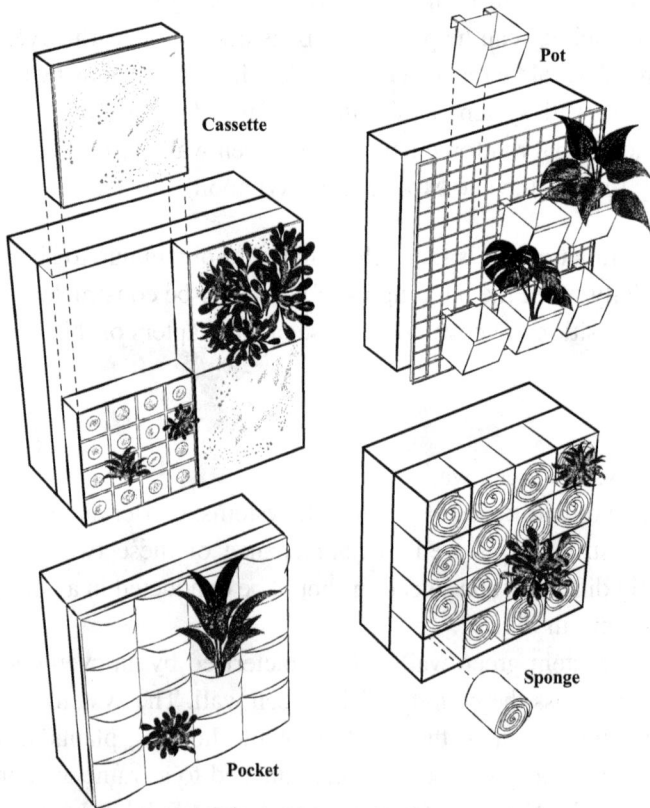

Fig. 26.   Common types of carrier system green walls.

design of the substructure is highly dependent on the size and location of the green wall. It is normally planned only after a visit to the site and with clear understanding of the loading requirements of the entire system.

Some substructure designs provide sufficient storage space for mechanical equipment such as pumps and timers, as well as sensors and data loggers for monitoring green wall conditions.

b. Carrier. The carriers can be mounted after the substructure has been securely installed onto the existing wall. These carriers are essentially containers for substrate and plants. Some examples include modular pots with hooks, cassettes with regularly-spaced cavities, and foam panels layered with moisture-absorbing fabric with pockets for holding plants (Fig. 26). These variations provide different advantages for the green wall, such as the ease of replacement of individual plants for modular pot carrier systems. The exact method of fixing the carrier onto the substructure depends on the system in use. Some systems provide additional fasteners for improved stability.

c. Plant and substrate. These two components will be housed inside the carrier. Depending on the size and orientation of the carrier, different plants and substrates will be used. Most carrier green wall systems do not support plants that are too big, as there is a limitation to the size of the root ball that can be contained within the carrier.

d. Irrigation system. Due to the extensive height and coverage of green walls, regular and evenly distributed irrigation needs to be provided to ensure plant survival. The irrigation system typically consists of a water source (faucet or water tank), pipes that cover the entire green wall at intervals, and designated drainage. This can be the existing drainage for outdoor green walls or a tray that directs water back into the water tank for indoor green walls. Irrigation pipes are normally fastened onto the substructure metal mesh using cable ties. A generous number of emitters are attached onto the pipes to ensure that water flows directly into the substrate that is close to the plant roots. For systems that require nutrient distribution, a separate fertigation tank is normally installed with the water tank. For such systems, regular checks should be conducted to ensure that both water and liquid fertilisers are not depleted.

Since carrier green walls are proprietary, product information can be readily found from respective vendors. While it is not necessary for designers to acquire in-depth knowledge of carrier green wall systems prior to implementing them into their design (since it comes as an entire package), some key considerations for selecting this particular system include the following:

i. Function. Designers have to decide on the main purpose for wanting to erect a green wall in the first place. They can be for aesthetics (e.g., to decorate the lobby of a hotel so that it looks more inviting to guests) or to serve a specific function (e.g., to increase the greenery density in order to fulfil prevailing urban greening policies). The appropriate green wall system should be recommended in view of its function. For instance, carrier system green walls are better suited for ornamental purposes because the carriers provide for a larger plant palette, compared to support systems that rely chiefly on climber plants. In general, carrier system green walls are ideal when precision in plant placement and composition are desired. Replacement of plants is more convenient as specific carriers can be identified and swapped with a substitute, thus maintaining the aesthetic quality of the green wall.

ii. Loading. For most carrier systems, the load of the carrier and plants is laterally transferred to the superstructure (existing wall) through the substructure. Therefore, the existing wall must be able to withstand the additional loading. In general, if the existing wall can be retrofitted with a new cladding, it should be able to handle the load of a carrier green wall system (subject to assessment from qualified personnel). Carrier systems may be harder to install on glazed façades compared to concrete walls, as the substructure has to be securely fastened onto the wall. Loading can differ significantly between carrier systems, such as cassette and foam systems.

iii. Transparency and access. Carrier green wall systems consist of several opaque layers, making them unsuitable for glazed façades or walls with large openable windows, where unobstructed views are desired. Care should also be taken to avoid placing green walls at places that may obstruct fire escape windows.

iv. Plant selection. Some basic requirements for plant selection include lighting requirements and carrier size. Although carrier green wall systems have a larger plant palette, designers need not necessarily opt for a high number of species to be planted. Instead, plant selection should preferably be based on the function of the green wall. If the purpose is mainly decorative and the green wall is to be sited at a location of prominence, designers may consider introducing a larger variety of plant species comprising different leaf sizes and colours, as well as an aesthetically pleasing composition for the entire green wall. If the green wall is to be placed in an obscure location with little traffic, it may be reasonable to reduce the plant variety and focus on maintainability of the system.

Different carrier systems also have different carrier sizes and methods for holding plants, which further differentiates plant selection. Cassette systems have their openings facing sideways, compared to pot systems that have upward-facing and much bigger openings for each plant.

Plant selection should also be considered in tandem with the design composition of the entire green wall. Due to the size, high growth rate and invasive nature of some plant species, they may end up encroaching and covering smaller adjacent plants that are in close proximity. This interaction between plants over time may lead to an overall effect that is drastically different from the initial proposed design. For improving maintainability, plants that grow too fast or too large over time should be avoided.

v. Location. Carrier system green walls are designed for the carriers to be replaced as and when necessary. Although the process of replacement is straightforward, it may become challenging in practice, with issues such as a lack of safety access or extreme height. When designing a green wall, its location should be critically examined for ease of access and safety during maintenance. If it is too costly or dangerous to reach every single pot (e.g., cherry pickers have to be rented to reach higher pots), it would be more feasible to consider support system green walls as an alternative.

vi. Cost. The large number of components and structural requirements of carrier system green walls inevitably leads to higher installation and operation costs compared to support system green walls. The difference in costs becomes exceedingly clear as wall area increases. Where adoption of carrier system green walls is still tenable, high costs can be mitigated by clarifying the function of the green wall, as well as careful selection of plants to minimise maintenance.

2. Support

Plants in support systems grow from the ground, vertically upwards against the walls, or with the help of support systems such as a trellis (Perini, Ottelé, Haas, & Raiteri, 2011). Support system green walls have their substrate only at one end of the green wall. Typically, the substrate is only provided at the bottom of the green wall. Plants are expected to grow upwards with the help of physical supports and provide coverage through their foliage over time. The system is designed with modular support components that begin at the substrate and end at the designated green wall area. The main parts of support green walls are as follows (Fig. 27):

a. Planter area. Most support system green walls use planter boxes, which can be raised from the floor or sunken to create a flushed appearance with the finished floor level. Raised plant boxes can be constructed with concrete and masonry and subsequently filled with soil. Alternatively, large pots can also be used.

b. Support system. This component will occupy the bulk of the green wall area. The function of the support system is to provide physical support for climber plants to attach themselves securely and to carry the weight of the plants without compromising the structural integrity of the green wall or existing wall. Typically, the support will start at the planter area and is located very close to the plants to facilitate climbing. The support system itself needs to be fastened onto a substructure or existing wall. However, it need not be as elaborate as carrier system green walls as its loading is significantly lower. Support systems can take on many forms, such as steel cables, wooden trellises, metal meshes, etc. As long as adequate support is provided, there is no restriction to the design and material of the support system.

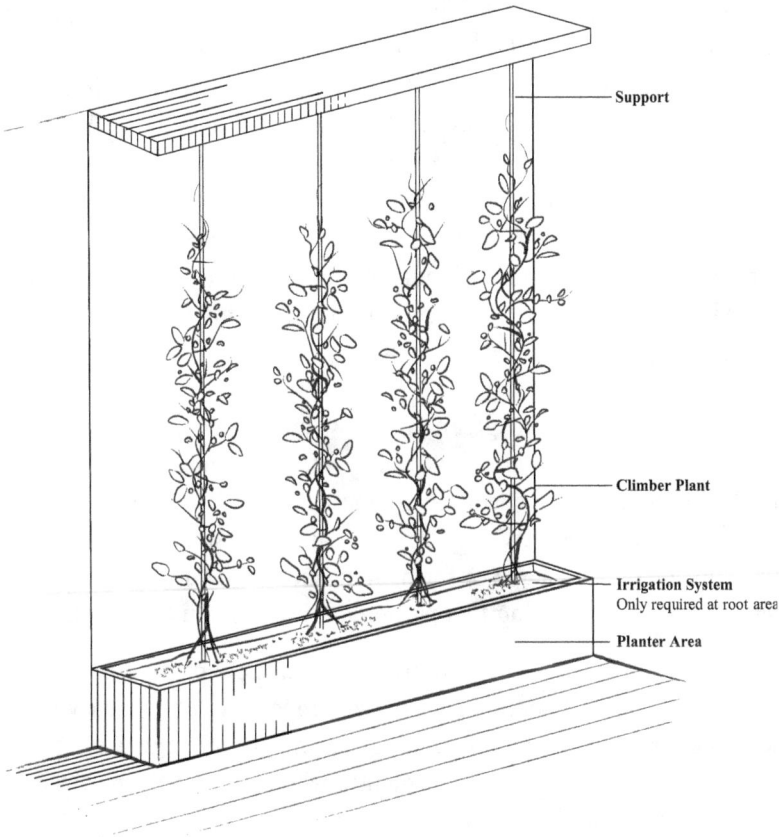

Fig. 27.   Components of support system green walls.

c. Plant and substrate. For support system green walls, only climber plants can be used. Climber plants are defined as plants that climb up trees or other tall objects. Plants have differing climbing mechanisms as well as climbing rates. There are two board categories for climber plants, namely, (i) Vines that attach themselves to support by making use of tendrils, twining petioles, adhesive pads, thorns, hooks or clinging roots, and (ii) vines that twine their stems around a support (Fiorello, Del Dottore, Tramacere, & Mazzolai, 2020). Due to the size of planter areas, a combination of top soil and coarser aggregates is normally used to fill up the planter boxes.

d. Irrigation system. As the planter area is located on one end of the support system green wall, irrigation is provided only for this area. There is no need to provide irrigation on the support system, as water is absorbed through the roots of the plants, which are in the planter boxes. This is significantly different from carrier system green walls, where irrigation has to be provided for the entire surface area of the green wall. In addition to irrigation, drainage for planter boxes have to be provided. Over time, compacted soil may result in poor drainage, leading to flooding in the planter boxes during periods of prolonged rain. This may cause root rot and eventual plant death.

Unlike carrier system green walls, where the design is proprietary, support system green walls have many more rudimentary components with details that can be specified by the designer. Therefore, as long as basic structural and safety requirements are met, support system green walls can take on any shape or form. With such high variability, some key considerations for selecting this particular system include the following:

i. Function. The purpose of wanting to have a support system green wall should be clarified. Generally speaking, support system green walls are less aesthetically pleasing compared to carrier system green walls because (i) immediately after installation, plant coverage will only be limited to areas close to the planter area, as plants take time to grow and cover the entirety of the green wall, (ii) only climber species plants can be used, resulting in less variety in plant leaf colour and size and (iii) the composition of the green wall cannot be varied for creative purposes, due to the limited plant palette as well as a common planting area.

Support system green walls are ideal for large-scale coverage to improve overall greenery density, as planter boxes and support systems are cost-effective and do not require much maintenance with proper design and construction.

ii. Loading. The weight of support systems may vary significantly: For cable-based systems, the overall weight consists of steel cables and anchors. For wooden trellises and metal meshes, additional weight from substructures is expected. Overall loading of the support system on the existing wall should be assessed by qualified personnel.

iii. Transparency and access. Some supports tend to consist of thin members (steel cables are 5 mm to 10 mm in diameter) and are spaced widely apart, making such systems suitable for installation on glazed façades or walls with large fenestrations.

iv. Support type. The common forms of support systems are cable, trellis and mesh systems. These systems provide physical support for climber plants to attach themselves firmly whilst climbing upwards (sometimes downwards). When selecting the support type, it is vital to do so in consideration of the plant species to be planted for the green wall. Different species of climber plants adopt different climbing strategies, and the support type must complement the specific climbing method of the plant it is supporting (Fig. 28). This involves an assessment of the material of the support and its alignment, as well as spacing between supports.

v. Plant selection. For support system green walls, plant selection is limited to climber plants (upwards) or creeper plants (downwards) only. Plant selection has to be conducted in tandem with a suitable support type. Flowering climber plants such as *Thunbergia grandiflora* and *Bauhinia kockiana* can be used for improved aesthetics and biodiversity. Besides plant species, the number of plants allocated to each support member should be specified. It is not necessary to allocate one support per plant: multiple plants can use the same support. This can provide better support between plants as they grow and provide lusher foliage for the green wall.

vi. Coverage. This is an important consideration that is often overlooked. During the design stage, green walls are often wildly romanticised to be fully covered with lush, sparkling foliage and held up by structures that are depicted as light and inconspicuous. In reality, and especially for support system green walls, initial coverage is extremely low, and the eventual coverage tends to be unevenly distributed (Fig. 29). Initial coverage is low because plants need time to climb up the support, and that can take months, if not years. Eventual coverage can be uneven owing to a myriad of factors, such as the spacing between supports, which leaves gaps between climbers, clustering of plants nearer to the bottom or top of the green wall, or external factors that hinder the growth of plants. Aggressive climber plants may even grow beyond their designated area and encroach onto other parts of the building.

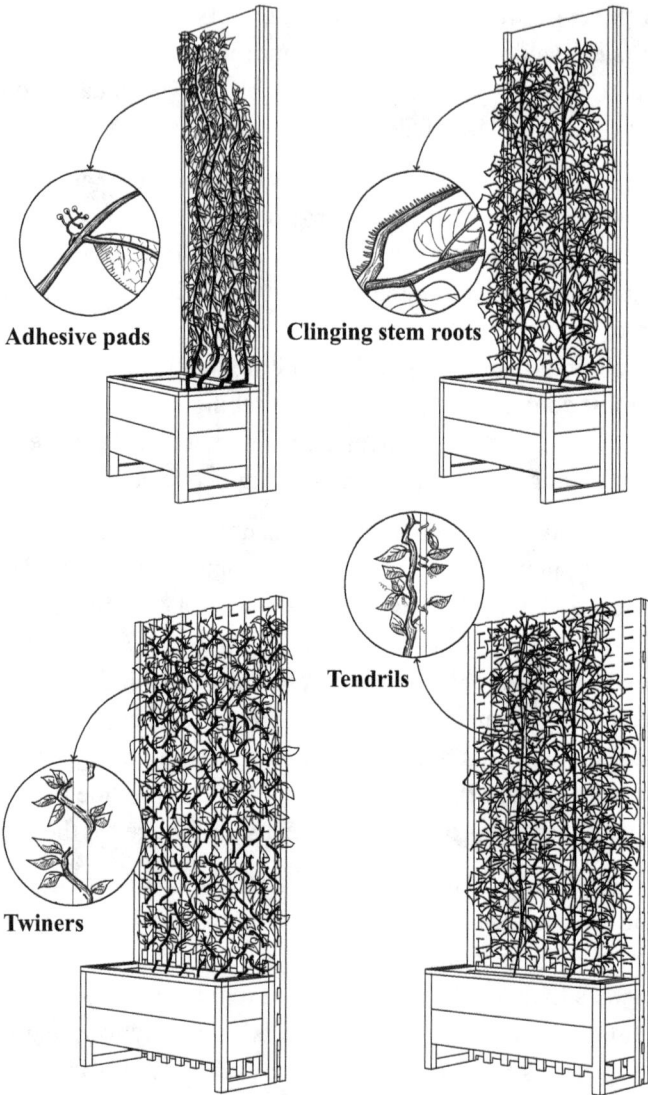

Fig. 28.   Types of climbing strategies for plants and support systems.

vii. Location. Besides the basic requirements of sunlight and water, support system green walls should also be placed at locations where maintenance can be conducted without hindrance. For excessively high green walls, it will be necessary to compartmentalise them into smaller

Fig. 29. Uneven coverage for a support system green wall.

sections, with planter boxes allocated at intermittent levels. For instance, a 10-storey-high support system green wall can be designed as five 2-storey-high green walls with planter boxes at each storey where the new green wall starts. In this manner, plants need to only climb up 2 storeys instead of 10. The natural growing limit of plants should be acknowledged.

viii. Cost. Compared to carrier system green walls, support system green walls are much more cost-effective. There are fewer components and plant numbers. Typically, the need for maintenance is much lower for support system green walls once they are established.

Plant

1. Health

Plants are the most important and prominent aspect of green walls, and selection should be considered in conjunction with environmental as well as system requirements. When selecting plants for green walls, the priority is to keep them alive and in good health. In this context, good health is defined as being in a perpetual state of well-being, growing at a normal rate and being free from pests or diseases.

Only when plants are in good health and experiencing normal growth can relevant ecosystem services be delivered as intended. For instance, if plants are under water stress, leaves might wither. Foliage density will decrease and shade provision (thermal regulation as the intended form of ecosystem service) will be reduced. Plants in good health can also reduce their chances of being attacked by pests and diseases.

When plants are in poor health, there will be marked differences in their outward appearance, such as foliage colour and canopy structure, some of which may become safety hazards (e.g., falling branches). There is a high probability that plants in poor health may eventually die and will need to be replaced, increasing overall maintenance and nullifying any benefits that green walls may bring.

To ensure that plants on green walls can be in good health, it is necessary to understand basic plant physiological requirements and ensure that those requirements are met. This is covered in subsequent chapters of this book. In general, plants will require sufficient light and water to grow. They should be planted in a good substrate that is sufficiently porous and provides the proper nutrients. The area of planting should be free of pests and diseases.

For light provision, plant selection can be based on understanding the microclimate (i.e., lighting conditions onsite), thereafter selecting plants that match the prevailing lighting conditions. For instance, if the site is perpetually shaded, designers may opt for shade plants, such as ferns instead of plants that require longer duration of direct sunlight exposure such as *Coleus scutellarioides*.

Irrigation requirements of different plant species should also be considered during selection. This can be assessed in tandem with the substrate mixture, which can provide some indication of the porosity and water-retaining capabilities of the substrate.

Such information can help determine if the surrounding environment, as well as the green wall system, can support plant life. They are vital for making sensible choices that can preserve the integrity of the green wall over time and minimise costs associated with the maintenance and replacement of plants.

Information on light and water requirements for plants can be obtained from several sources, such as books on gardens and plant care and research journal publications or from online sources such as *Flora & Fauna Web* (https://www.nparks.gov.sg/florafaunaweb), a website managed by the National Parks Board of Singapore. Common information, such as lighting preference, water preference and growth rate, can be accessed via the website.

2. Maintainability

Maintainability associated with plants constitutes the next major consideration. This is exclusive to plants and is independent of any green wall system components (covered in previous sections).

Unlike considerations for plant health, where the objective is to ensure that plants are in a perpetual state of well-being and are thriving, the focus here is on preserving the state of plants in a manner where further intervention can be minimised. Intervention occurs mainly in two ways: when plants are not growing well and wither, maintenance is required for replacement, and when plants are growing too well, maintenance is required for pruning.

Therefore, plant consideration for health has to be conducted in tandem with maintainability. In practical terms, it would entail choosing plants so that they are able to survive in the ambient environment but will not grow too quickly or large to subject the green wall to issues of maintenance or safety.

Maintainability for green wall plants is a subject that is often overlooked, as plant selection is not a prominent phase in the green wall commissioning process. Designers often pay more attention to the system that is used, operating under the assumption that the wall will be "green with leaves" once it is up, and leaving it at that. When dealing with carrier systems, planting composition takes precedence over species selection (i.e., priority is given to visual impact, such as forming waveforms using plants rather than what plants species are selected). Plant selection is often proposed by the carrier system green wall supplier, together with the proposed plant composition. Designers tend to accept the proposed plant palette due to a lack of knowledge of the specific plant species and assurances given by suppliers that the nominated plants are suitable for the system. For support system green walls, designers may consult landscape architects or landscaping companies for suitable plant choices since they may already be handling landscaping works for the project. In this case, the focus would be on ensuring adequate coverage for the green wall and not so much on the actual species. In some cases, more resources have to be allocated for maintenance, such as walkways to reach all parts of the green wall or using equipment such as gondolas or cherry pickers for regular inspection and maintenance works.

The key consideration to selecting plants for improving maintainability is to understand the growing traits of the plant.

For example, to minimise the pruning of plants on carrier system walls so that the design pattern can be preserved, plants that grow too quickly such as *Tradescantia zebrina* should be avoided. Large ferns such as *Davallia denticulata* may cover adjacent, smaller plants as they grow and lead to bald spots as the smaller plants wither. *Philodendron hederaceum* roots are invasive and will grow beyond pockets of fabric-based carrier systems, leading to the disintegration of the fabric and intertwining roots that make plant replacement difficult over time (Fig. 30). Using drought-tolerant plants may improve the resilience of carrier system green walls in terms of their irrigation requirements.

Plant selection for maintainability can be conducted in view of the function of the green wall: if the green wall is located in an area where it cannot be observed in detail or where visual exposure is low, designers may opt for plants that look less aesthetically pleasing but are easier to maintain. Plants such as *Tabernaemontana divaricata*, normally planted for their white pinwheel flowers, can be used when the green wall can be observed at close range but may not be necessary when the green wall can only be viewed from far away.

Green roofs

Green roofs refer to horizontal or inclined surfaces that are covered with plants to a large extent. They are a form of building envelope greening, typically for the building exterior. Roof gardens is another common name for green roofs. The concept of greening rooftops is not new, with the Hanging Gardens of Babylon (built around the 6th century B.C.) being a

Fig. 30.   Invasive roots on green wall fabric.

prominent example. These green roofs (or roof gardens) did not differ much from ground-level gardens, besides not having large trees due to loading limitations. Modern interpretations of the green roof can be found in the early 1960s, mainly in Europe. Since then, the usage of green roofs has propagated to many parts of the world (Shafique, Kim, & Rafiq, 2018). Sky terraces, which are installed in parts of the building other than the roof, are functionally similar to green roofs. As such, most of the information covered here on green roofs is also applicable to sky terraces.

Implementation framework

When deciding to have green roofs in a project, it is important to consider several key points to minimise failure. The three main considerations are to:

1. **Ascertain the benefits of having the green roof.** There are many reasons for building owners to want to install a green roof in their estate. Some may want to increase the greenery density of the building so that it will comply with building or developmental regulations. Green roofs can also provide thermal insulation to reduce heat gain into buildings through the roof surface. Having landscape features on the rooftop can help improve the aesthetics of the roofscape and make them more suitable for social activities. Similar to green walls, once a desired benefit is identified, it may be optimised. For instance, if temperature reduction is the desired benefit, intensive green roof systems may be installed, and plants with bigger foliage can be used for added thermal insulation. Benefits can be direct or indirect: a decrease in surface temperature can be observed directly, whereas savings in energy consumption is indirectly estimated from the net heat gain into the building and subsequent reduction in cooling energy.

2. **Tabulate the costs of having the green roof.** In addition to the initial costs of installing the green roof, maintenance costs may add up over its life span. The cost of installation and maintenance of green roof systems varies significantly. Costings and life cycle data on green roofs provide insight on methods for minimising costs, such as designing for ease and safety during maintenance. Just like benefits, costs can be direct (arising from installation) or indirect (plant replacement, labour from maintenance, etc.)

3. **Consider factors that will affect green roof installation and maintainability.** Some examples include green roof design, plant selection, maintenance regime, irrigation schedule, etc. Similar to the green wall assessment framework, these considerations are further broken down into environmental, system and plant considerations. These factors can significantly impact the overall value of the green roof.

The proposed framework for assessing the suitability of installing a green roof is shown in Fig. 31. The main criteria of Benefits, Costs and Considerations should be comprehensively evaluated before embarking on a green roof project.

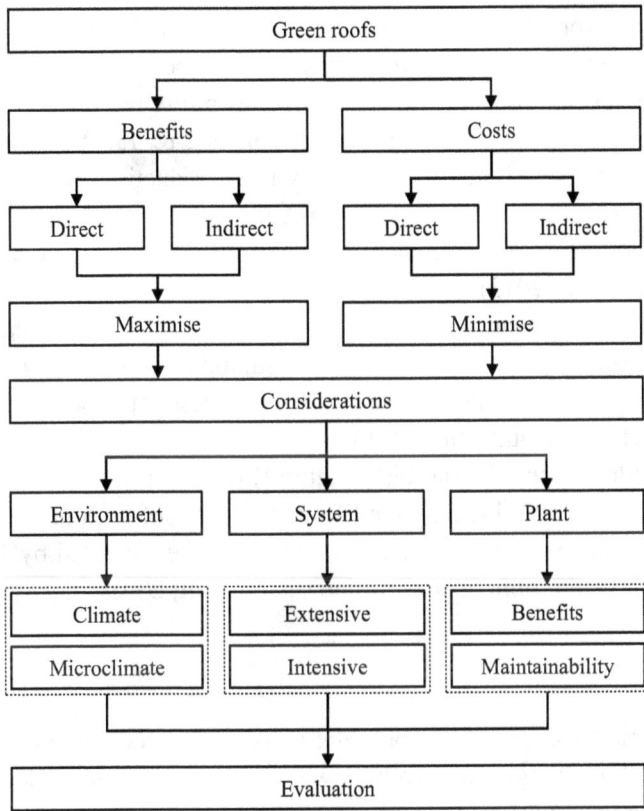

Fig. 31.   Implementation framework for green roofs.

Benefits

Temperature

Roof surfaces are exposed to high amounts of direct solar radiation, which leads to high surface temperatures. A portion of this heat gain will be transmitted into the building, through its roof, and into the interior surfaces. This can increase the temperature of the spaces under the roof, leading to thermal discomfort for occupants.

Numerous studies have shown how green roofs can reduce the temperature on roof surfaces. This is possible as green roofs typically consist of several layers that act as thermal insulation for the building surface. As such, the surface temperature of the roof can be significantly lowered. Some studies have documented surface temperature reduction of close to 30°C with green roofs (Wong, Tan, Kolokotsa, & Takebayashi, 2021). When the green roof area is sufficiently large, reductions in air temperature can also be observed. Large shrubs and trees on green roofs help cool the surrounding area via higher shade provision and evapotranspiration (Saadatian et al., 2013).

Since dense urban areas often lack ground-level space for large parcels of greenery, green roofs provide a suitable alternative by greening spaces above ground and addressing issues of rising temperatures due to the UHI effect (Santamouris, 2014).

By reducing the diurnal temperature fluctuation on the roof surface, green roofs can help increase the lifespan of the roof by limiting thermal stress. The lifespan of roofing membranes may be extended by 10 to 20 years with green roofs (Clark, Adriaens, & Talbot, 2008).

Energy

As heat transmission into the building is reduced in the presence of green roofs, cooling energy can also be lowered. This is because a decrease in ambient temperature indoors will lead to lesser energy required to cool the spaces. Savings in energy consumption due to the presence of green roofs have been observed in studies in different parts of the world, under varying climatic conditions (Pisello, Piselli, & Cotana, 2015; Tariku & Hagos, 2022; Wong, Tan, & Chen, 2007).

There is some indication from studies that cooling from green roofs is most significant during the hottest and driest conditions (Alexandri & Jones, 2008). This shows that the cooling potential of green roofs is not constant throughout the year and may be subjected to environmental or plant factors, such as the amount of foliage. If the intention of the designer is to tap on the energy-saving potential of green roofs, it is advisable to select plants that can provide more cooling and to site the green roofs above air-conditioned spaces to reduce heat transmission.

Carbon

A widely adopted phrase when addressing the impact of green roofs on carbon footprint is "carbon sequestration", which describes the process of capturing and storing atmospheric carbon dioxide. In the context of green roofs, this is possible by means of the plant's photosynthesis process, whereby CO_2 from the surroundings is captured (Velasco & Roth, 2010). In addition, the temperature reduction and energy savings potential of green roofs can further reduce CO_2 emissions from buildings (Sadineni, Madala, & Boehm, 2011).

For identifying the contributions of green roofs in terms of carbon sequestration, it is useful to categorise them into direct and indirect benefits. Since carbon sequestration involves the capture and storage of carbon in plants and substrate, green roofs contribute directly to this process through the vegetation and soil layers of the green roof system. To strengthen the sequestration potential of the green roof, suitable plants should be selected (Zhao, Tabares-Velasco, Srebric, Komarneni, & Berghage, 2014). The depth and moisture content of the substrate layer plays another important role in facilitating carbon sequestration (Kazemi & Mohorko, 2017).

The presence of green roofs can also provide carbon sequestration in an indirect manner. The combined effects of shade provision and evapotranspiration reduce net heat gain into the building, leading to lower energy consumption and lesser burning of fossil fuels, and, in turn, reducing CO_2 emission. Studies involving the simulation of building energy consumption have noted vast reductions in energy usage when rooftops are covered with green roofs (Berardi, GhaffarianHoseini, & GhaffarianHoseini, 2014; Mahmoud, Asif, Hassanain, Babsail, & Sanni-Anibire, 2017). Unlike green walls, green roofs have a much higher potential for carbon sequestration as the former can accommodate larger plants or even trees and typically hold a larger volume of substrate within the system.

Water

Green roofs can improve the hydrology of urban areas directly in two ways. Firstly, green roofs can improve stormwater retention capability to reduce runoff from building surfaces. Due to the prevalent usage of impervious materials such as glass, steel, tarmac and concrete, there is an increased chance of flooding in urban areas. One way of alleviating peak runoff is via on-site detention and retention, where rainwater that hits building surfaces can be stored temporarily in parts of the green roof such as the substrate or reservoir layer (PUB, 2014). The substrate of green roofs can absorb significant quantities of rainwater, which can also be used to irrigate plants. Many studies have looked into the water retention capability of green roofs in terms of substrate and plant selection. Results show green roofs to be effective in reducing peak flow and runoff in many parts of the world, regardless of the type of plant or substrate used (Bengtsson, 2005; Carter & Jackson, 2007; Mickovski, Buss, McKenzie, & Sökmener, 2013).

Secondly, green roofs may improve water quality by absorbing pollutants from rainwater. Studies have shown a reduction of heavy metal concentration in the runoff due to green roofs (Berndtsson, Emilsson, & Bengtsson, 2006). For benefits to be optimised, it is important to select the right green roof system, substrate, fertiliser and plant species. If not, the pollutant count may increase instead (Berndtsson, 2010).

Air quality

Air pollution is a common problem in highly urbanised or industrialised areas. Fine dust particles in the air may adversely affect the health and well-being of city dwellers. One manner in which greenery can help reduce air pollution is to capture the pollutants such as O_3, NO_2, SO_2, CO and dust particles through the plant leaves and stomata (Manso, Teotónio, Silva, & Cruz, 2021; Yang, Yu, & Gong, 2008). The impact can be substantial when applied on the scale of a large garden or park and extended to green roofs. Plant selection is an important factor in improving air quality, as plant functional traits such as leaf size and canopy structure can determine the overall effectiveness of dust capture (Shafique, Kim, & Rafiq, 2018). Should air purification be a desired benefit, intensive green roof systems should be installed, with larger shrubs and trees preferred.

Green roofs may also indirectly improve air quality via less carbon emission into the atmosphere from the reduced cooling load associated with lower heat transmission into buildings.

Acoustics

Similar to green walls, green roofs can help absorb sound through the plants. With green roofs, the substantial substrate layer offers added insulation from sound. This helps to reduce reverberation to the surroundings as well as transmission into buildings. A study reported attenuations of over 10 dB over extensive green roofs for frequencies of between 400 Hz and 1,250 Hz (Van Renterghem & Botteldooren, 2011).

Biodiversity

Urbanisation leads to habitat fragmentation, which compromises the conservation of flora and fauna biodiversity. Reduction of biodiversity may lead to the decimation of plants and animals. With lesser pollinator species, such as bees and birds, the propagation of green spaces may be impeded. This can also have an adverse impact on crop yield as well as the UHI effect, as the food provisionary and thermal regulatory benefits of greenery are curtailed.

Recognising the need for biodiversity preservation, many cities have begun to initiate policies that seek to protect or increase the size of greenery in the urban environment (Jim, 2004; Lidmo, Bogason, & Turunen, 2020). One common strategy is to promote the installation of green roofs so that previously untapped roof surfaces can have the added function of increasing greenery density (Carter & Fowler, 2008).

Several studies have shown how green roofs can reduce habitat loss for wildlife in urban areas. Encouraging signs of increased diversity and species count have been documented (MacIvor & Lundholm, 2011). When attempting to tap into the potential of green roofs for preserving and promoting biodiversity, it is useful to consider how appropriate plant selection can help to attract the right fauna. For instance, in order to attract bees, flowering plant species such as *Sphagneticola trilobata* and *Lantana camara* can be used.

Social and aesthetic quality

Of the plethora of ecosystem services that green roofs offer, perhaps none stands out more conspicuously than providing social benefits and improving the aesthetic quality of the roofscape.

As largely functional spaces, rooftops are typically used to house mechanical equipment such as HVAC chillers and cooling towers. Continuous exposure to high levels of solar irradiance and wind conditions means that these spaces are not conducive for social gatherings. With the introduction of landscape features, roof spaces can become comfortable and habitable. Thermal comfort can be improved with shade from trees, and strong winds can be blocked with tall shrubs. The combination of hardscape (e.g., benches) and softscape with green roofs can turn harsh and barren rooftops into lush gardens and recreational spaces. Green roofs can also be used for urban farming, to promote a culture of self-reliance for food production and foster better community spirit through the farming process.

The creation of these additional "habitable" spaces can often lead to an increase in property value. Many studies have observed that the commercial value of real estate tends to increase when in close proximity to green spaces (Belcher & Chisholm, 2018). In similar fashion, studies on the economic impact of green roofs have noted increases in property value (Bianchini & Hewage, 2012b), which can lead to direct monetary benefits, such as an increase in rental prices. In one study, an estimated 16.2% increase in rental prices for buildings with green roofs was reported (Ichihara & Cohen, 2011).

The presence of green roofs can also help foster notions of sustainability as well as environmentally friendly practices to its occupants. The presence of greenery can create the right ambience and serve as a reminder for active participation in energy-saving practices and the reduction of carbon footprint in our lifestyle.

Costs

The costs associated with green roofs include expenditure for the green roof system, plants, existing roofing works, irrigation, manpower and maintenance. The unit cost of green roofs may be lower than that compared to green walls, but due to the scale of green roof installations, the overall cost may still be substantial. It is therefore important to be aware of the costs involved and how they can be minimised.

Pre-installation and installation

Pre-installation costs refer to costs associated with the green roof before construction. This may include roofing works, such as the application of a waterproof membrane or replacement of roof tiles, to ensure that there are no water leakage issues after the green roof is installed. Some green roof vendors provide pre-installation services as part of the purchase of the green roof. Designers need to ensure that effective irrigation and drainage are provided prior to installation, as any rectification after the green roof has been installed may result in a large-scale overhaul and exorbitant costs. Extensive green roof systems are suitable for retrofit on rooftops, but intensive green roof systems may require professional assessment for loading and structural integrity due to the large soil volume as well as trees. Additional loading due to intensive green roofs can be up to 30 kN/m^2. The structural costs associated with intensive roofs can be significant: Wong, Tay, Wong, Ong, and Sia (2003) estimate the structural cost of a roof deck with an intensive green roof to be about 50% more than a roof deck without a green roof installed (SGD 32/m^2 more).

The installation of green roofs takes into consideration items such as plants, substrate, green roof system, transportation and manpower. Here, installation costs refer to:

1. Preparing the site to be suitable for green roof installation.
2. Installing the green roof system.
3. Installing auxiliary components of the green roof, such as an irrigation system (pumps, pipes and emitters) or soil moisture sensors.
4. Supply and installation of plants and substrate on the green roof.
5. Obtaining approval from authorities (such as PE certification for loading, etc.).

The indicative costs of green roof installation can vary significantly due to differences in system type, size, locale and maturity of the industry. In one study, the average cost of extensive green roof systems was USD 112/m^2, and the average cost of intensive green roof systems was USD 409/m^2 (Manso, Teotónio, Silva, & Cruz, 2021). In Singapore, prices of green roofs range from SGD 40/m^2 to SGD 65/m^2, depending on the green roof system and the proportion of shrubs and trees used. In general, higher costs are associated with intensive green roofs as better water-proofing material, more soil and bigger plants (e.g., trees) are used (Wong, Tay, Wong, Ong, & Sia, 2003). Typically, the price of green roofs tends to decrease as the industry matures and there are more vendors and products to choose from (Feng & Hewage, 2018).

Maintenance and disposal

Maintenance of green roofs constitutes a major portion of their Life Cycle Cost (LCC). Items that normally require maintenance are (i) the green roof system (hardware), (ii) irrigation system, and (iii) plants. Green roof systems, if left undisturbed, do not require much maintenance. Most systems are designed for outdoor use and are made of durable materials such as polypropylene. Irrigation systems may require more attention, as pipes or tap fittings may be dislodged or disintegrate over time. When located near areas with large parks, irrigation pipes may be damaged by wildlife (chewed by squirrels). Compromised irrigation will result in leaks, causing water penetration to unintended areas and reduced water intake for plants. If left unfixed, plants may eventually wither. The final item that requires the most maintenance is the plants. Manpower is required for regular inspection and pruning to ensure that plants do not grow beyond their designated areas. For bigger plants such as trees, arborists may need to be engaged to ensure that they are in good health and will not pose any safety hazards (e.g., falling branches). Plant replacement may also be required if growing conditions are unsuitable and plants start to exhibit signs of stress or cellular damage. Pests and diseases may also be spread across plants, resulting in the need for additional treatment or replacement. Details on site assessment methods to determine plant suitability are covered in the next chapter.

These factors, together with the prevailing climate, will determine the overall maintainability of the green roof. More attention should be paid during the initial weeks of installation, to ensure that plants can adapt well under the new ambient conditions. Studies on the costing of green roof maintenance costs show a range of between USD $4.84/m^2$/year and USD $6.37/m^2$/year (Manso Teotónio, Silva, & Cruz, 2021). The cost of disposal for green roofs include the plant, substrate and green roof system removal, transport and taxes (Peri, Traverso, Finkbeiner, & Rizzo, 2012). Reinstatement of the roof surface may be required in some cases. The cost of disposal has been documented to be in the range of USD 14/m² to USD 29/m² (Bianchini & Hewage, 2012b; Perini & Rosasco, 2016; Sproul, Wan, Mandel, & Rosenfeld, 2014).

Considerations

There are three main factors to consider before commissioning a green roof project. Similar to green walls, they are categorised into (i) environmental, (ii) system, and (iii) plant considerations (Fig. 31). These factors may impact the performance of green roofs on their own or through interaction over time to cause significant change to overall quality. There is no order of importance for the categories.

These recommendations are based on fieldwork experience, interviews with industry experts, as well as reviewed literature.

Environment

1. Climate
Climate refers to prevailing conditions where the green roof is to be installed. It dictates the choice of plants suitable for the green roof, irrigation requirements, extraction of ecosystem services and the overall feasibility of having a green roof in the first place. Just like for green walls, important parameters include solar irradiance, rainfall, air temperature and relative humidity.

2. Microclimate
Information on the microclimate will help in plant and system selection to ensure the longevity of the green roof. The climate of a site and its microclimate can be vastly different, as the former is representative of prevailing weather patterns for a large area (which the site is a small part of), and the latter reflects actual conditions that are influenced by built morphology or anthropogenic activity.

For instance, climatic data from weather stations can show average wind speeds to be 3 m/s for a given precinct, but actual wind speed at the rooftop may be 0.5 m/s due to blockage from the building parapet.

Similarly, solar irradiance exposure may be high for tropical regions, but actual exposure can be much lower due to overshadowing from taller nearby buildings. Therefore, microclimatic data is important in determining local ambient conditions for making accurate decisions in green roof system and plant selection.

Sky terraces, which are landscaped spaces located at intermediate storeys of a building, exhibit more variability in microclimatic conditions as they are highly susceptible to the effects of self-shading and building orientation. The microclimatic conditions in these areas should not be assumed to be the same as ground level or rooftops.

Similarly, hybrid systems consisting of photovoltaic (PV) panels and green roofs underneath will have a heterogeneous lighting distribution under the panels, with low light access further inside the PV array and higher access along its periphery. Exact lighting conditions will depend on the height and arrangement of the PV panels. This will be exceedingly different from climate data obtained for the site. There will also be less access to rainwater as the PV panels will be sheltering the plants from rain. Where possible, field measurement or simulation should be conducted to ascertain conditions before deployment.

Microclimatic data for green roofs typically consists of the following parameters:

a. **Solar irradiance.** Microclimatic measurements of solar irradiance can help detect occurrences of self-shading and overshadowing. For hybrid systems or sky terraces, it can help quantify daylighting for assisting in the plant selection process or how they should be arranged to maximise light provision.

b. **Rainfall.** Rooftops tend to receive a high amount of rainfall due to their large surface area and orientation to the sky. Rainwater can be channelled into green roof systems with water retention reservoirs so that they can be used to ensure a consistent supply of water for plants (Tan et al., 2017). For areas such as sky terraces, rain may be partially (or even totally) blocked by nearby structures, resulting in low rainfall for the landscaped area. For hybrid systems, rain may be blocked by PV panels.

c. **Air temperature and relative humidity.** Provides information on ambient conditions, such as context-specific data like the impact of anthropogenic heat sources such as hot air from cooling towers or other mechanical services typically located on a rooftop, which may affect plant selection.

d. **Wind.** Wind profiles are seasonal and are highly influenced by urban parameters. For green roofs and sky terraces, higher winds can be expected due to increased elevation. In-situ factors can also affect wind exposure: Parapet and mechanical services can help to block wind, and eddy currents and the Venturi Effect may drastically increase wind speed to the point where it may affect green roof installation, especially if the roof is inclined. Therefore, measurement of wind speed and direction at the green roof installation site can help in both the plant and system selection process.

Both fieldwork and simulation-based data acquisition methods for the microclimate are covered in the next chapter.

System

Due to their potentially immense loading and structural impact on buildings, the type of green roof system must be carefully selected. Some initial considerations for having green roofs include the following (Fig. 32):

a. **Structural capacity of the roof.** The existing roof must be able to accommodate additional loads from the green roof, consisting largely of foliage, substrate and stormwater storage capacity. The actual loading is dependent on the type of green roof system selected. It is recommended to engage an architect, structural engineer or other qualified personnel to ensure that structural integrity of the building will not be compromised with the addition of a green roof.

b. **Pitch of the roof.** A flat roof with a very slight pitch (1 to 2%) for drainage is ideal for maximising green roof storage. For sloped roofs, installation of green roofs is still possible, but the appropriate system and plants need to be selected. Additional reinforcements, such as

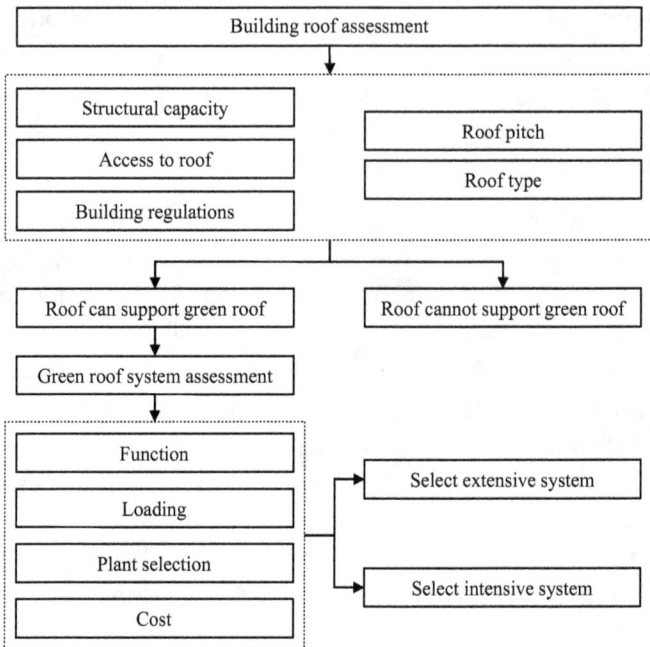

Fig. 32. Green roof system selection framework.

meshes and baffles, may be required to prevent any slippage of the substrate.

c. **Access to the roof.** Due to the large amount of construction materials, plants and organic matter required to install a green roof, it is necessary to consider how they can be transported safely onto the rooftop. Installation, storage and maintenance need to be planned for the sufficient area and safe access, in view of site conditions such as building height and other mechanical systems (e.g., chillers, PV panels) that will occupy the roof space.

d. **Type of roof.** Most building roofs will be able to support the installation of green roofs, except for glass roofs. Vegetation mats, mostly consisting of pre-cultivated sedum-moss plants, are light enough to be installed on metal roofs of bus stops (Tan, 2022).

e. **Prevailing building regulations.** Prior to installation, it is necessary to ensure compliance with all related building regulations and obtain all necessary clearances from relevant authorities. This may involve loading calculations, providing sufficient setbacks of green roof area for fire safety, submitting a maintenance plan, or estimating drainage capacity. Assessment should only be done by qualified personnel.

Intensive green roof systems

There are two main types of green roof systems: intensive and extensive. These two systems cater to different roof types and are different in areas of design, construction, usage and maintenance.

Intensive green roofs are characterised by their deep substrate depth (more than 20 cm). This provides a large plant palette that may include turf, shrubs and even trees, as well as fewer limitations to design variety. With the large volume of substrate as well as large plants, there is a considerable increase in loading. Most intensive green roofs can support live loads (i.e., pedestrian traffic) and can be landscaped to look like ground-level gardens, complete with shrubbery and trees, as well as furniture such as benches and gazebos. They are designed to be accessible and can be used for recreational and social activities.

The main components of intensive green roofs consist of the following (Fig. 33):

1. **Deck layer.** The roof deck supports the entire green roof system. Some common materials include concrete, metal, wood, plastic and slate. The roof itself will consist of several layers, such as insulation and waterproofing, depending on the main material used. The deck material will determine its load-bearing capacity and the composition of the intensive green roof.

2. **Waterproofing layer.** A high amount of moisture can be expected from the planned irrigation, as well as rainfall, stored in the substrate and plants. To prevent leakage from the green roof and any water damage, a waterproofing layer needs to be applied to cover the deck layer. There is a wide variety of waterproofing materials available, such as liquid-applied membranes, thermoplastic membranes and modified bitumen sheets (Townshend & Duggie, 2007). Care should be taken during the application of the waterproofing layer to ensure that there are no accidental gaps and that the waterproofing is protected from

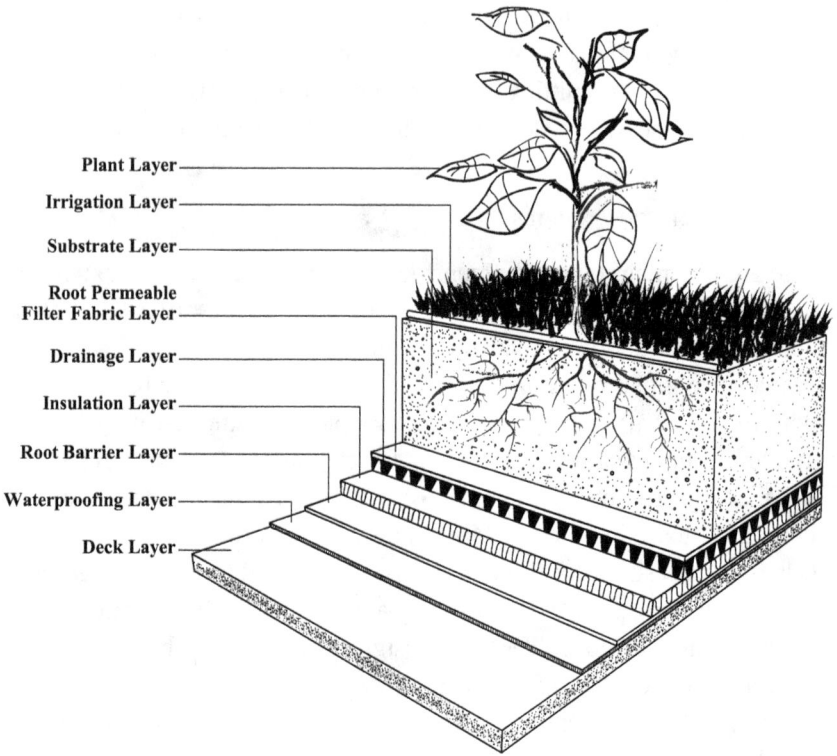

Fig. 33.   Typical components of intensive green roof systems.

chemical or physical damage. The lifespan of the waterproofing layer should be as long as, if not longer than, the components of the green roof system.

3. **Root barrier layer.** Plants with aggressive root systems might grow beyond the confines of the substrate space and result in root penetration through the waterproofing layer, causing cracks and water seepage into the deck layer. This will eventually lead to water damage of the deck layer. In some cases, roots can grow out of the green roof boundary and compromise the structural integrity of the green roof system (Bianchini & Hewage, 2012a). Therefore, root barriers such as polyethylene sheets are installed to protect the underlying layer from root penetration. Care should be taken to select materials that are chemically stable to minimise any chances of leaching or damage to plants.

4. **Insulation layer (Optional).** In temperate regions, having an insulation layer provides more thermal protection from extreme temperatures due to seasonal changes. There is reduced thermal transmission into the building, which may reduce cooling energy. It can also help provide added protection against physical damage and water seepage when installed above the waterproofing layer.

5. **Drainage layer and drainage system.** Effective drainage is crucial for preserving the integrity of the waterproofing layer and the entire green roof system. Poor drainage may lead to saturated substrate with reduced air content. This will adversely impact the rhizome and lead to root rot. The drainage layer can comprise natural aggregates (such as stone) or synthetic drainage material, such as looped polyamide filaments or polyethylene tray panels. Tray panels have small reservoirs that provide the added function of moisture retention for the substrate. The drainage layer will cover the entire area of the green roof footprint.

   The drainage system is another major component of the green roof. It collects excess irrigation and stormwater and directs them to the main roof drainage system. A common issue encountered by facility mangers is for the substrate to be compacted over time to an extent that the drainage point fails. This will result in saturated substrate and water overflowing from the planter box. To minimise clogging, it is recommended that coarser aggregates (larger stones) be placed around the drainage point and that drains be regularly inspected.

6. **Root-permeable filter fabric layer**. This layer is placed between the substrate layer and drainage layer to prevent loose growing media from clogging the drainage layer. It is semi-permeable to ensure the downward movement of water from the substrate into the drainage layer. Typically, a root-permeable polypropylene fabric layer is used.

7. **Substrate layer.** Also known as growing media, the substrate layer for intensive green roof systems can range from top soil to special soil mixes to enhance porosity and reduce overall weight. This may include components such as sand, pumice, vermiculite, peat, loam, lightweight expanded clay aggregates, etc. The exact mixture and depth of the substrate layer are dependent on the type of plants selected

for the intensive green roof. Bigger plants, such as trees, will require deeper substrate for better anchorage. Fertilisers will be applied at this layer.

8. **Irrigation layer.** A typical irrigation layer consists of pipes distributed evenly throughout the green roof area. Emitters are placed intermittently along the polyethylene pipes for drip irrigation. An irrigation control valve sits between the main pipe and water point. This normally comprises a battery-operated valve with a timer function to set the irrigation schedule. Advanced irrigation systems may also include soil moisture sensors to trigger irrigation when the soil moisture level decreases beyond a pre-determined threshold. One common cause of green roof failure is for plants to die due to irrigation failure. This is mainly due to damaged pipes, either from prolonged exposure to the elements or damage by animals (e.g., squirrels). Damaged pipes may lead to a loss in water pressure as well as leakages, leading to uneven distribution of water to the green roof. Another common occurrence is for timers to stop working after the batteries run out. Since the valves are normally hidden from view, it may remain undetected. Regular inspection of the irrigation layer is recommended to ensure that pipes and timers are in working order.

9. **Plant layer.** The final and most visible layer of the intensive green wall system is the plant layer. For intensive green roof systems, the choice of plants can range from turf to small trees. Selection of plants should include consideration of safety, loading and substrate depth. In general, larger plants require a deeper substrate. Plant selection can also include tapping onto specific ecosystem services, such as selecting edible plants for food, and promoting social activities, such as urban farming communities, or choosing flowering plants to attract butterflies and bees to promote biodiversity. Shade provision from trees can be a relevant ecosystem service (thermal regulation), but the size, loading and maintenance of big trees will also have to be factored into consideration.

Most intensive green roof systems are not proprietary. As shown in the previous section, there are many components to this system, and each component has its own range of options to choose from. For instance, applications for the waterproofing layer can range from liquid-applied rubber and elastomeric urethanes to form a seamless membrane, to membrane rolls that are laid out and mechanically fastened to the roof. Designers should select the components best suited for the green roof project by weighing the pros and cons of each option, as well as their associated costs.

Due to increasing interest in greening up urban areas, information on green roof design and installation techniques is readily available. There is ample literature on green roof construction, both in research as well as in practice. Some recommended references are presented in Table 7.

Some key considerations for selecting intensive green roof systems include the following:

1. **Function.** The reason for wanting to install a green roof should be clarified. If the purpose of having a green roof is to increase the greenery density, or public access is not intended, it may not be necessary to install an intensive green roof. Extensive green roofs offer comparable foliage cover without the need for much structural support. If the intention is to create a space for human occupancy in a social setting, then intensive green roofs may be an appropriate solution. The green roof can support live loads, people can walk on the turf, trees can be planted to provide shade, and the area can be made more conducive for activities.

Table 7.   References for green roof design and construction.

Topic	Reference
Green roof components	Luckett, Kelly. *Green Roof Construction and Maintenance.* McGraw-Hill Education, 2009.
	Weiler, Susan, and Katrin Scholz-Barth. *Green Roof Systems: A Guide to the Planning, Design, and Construction of Landscapes over Structure.* John Wiley & Sons, 2009.
Green roof maintenance	NParks CUGE Standards CS E07:2012
Design loads for green roofs	NParks CUGE Standards CS E01:2010
	NParks CUGE Standards CS E10:2014

(Continued)

Table 7. (*Continued*)

Topic	Reference
Filter, drainage and root penetration barrier layers for green roofs	NParks CUGE Standards CS E04:2010
Large planting for green roofs	NParks CUGE Standards CS E09:2012
Safety design for green roofs	NParks CUGE Standards CS E02:2010
Irrigation for green roofs	NParks CUGE Standards CS E06:2012
Pitched green roofs	NParks CUGE Standards CS E08:2012
Substrates for green roofs	NParks CUGE Standards CS E03:2010
Plants for green roofs	Tan, P. Y., and Angelia Sia. *A Selection of Plants for Green Roofs in Singapore*. National Parks Board, Singapore, 2008. NParks CUGE Standards CS E12:2017
Waterproofing for green roofs	NParks CUGE Standards CS E05:2012

2. **Loading.** Intensive green roof systems can be very heavy and require adequate structural support. This may pose a serious issue for retrofit projects where the building is old or the roof has not been designed to support significant additions in weight. This is a crucial point that may result in extremely adverse consequences if overlooked. Table 8 shows some examples of catastrophic green roof failures that have occurred, chiefly due to underestimation of the green roof loading. Loading can be broken down into dead loads and live loads: Dead loads are

Table 8. Examples of green roof failure. (Details can be found on www.ugl.sg/book.)

Location	Year	Cause of failure	Extent of damage
Chicago	2011	Overloading (snow)	Roof collapsed
Latvia	2013	Overloading (rain)	Roof collapsed, fatalities
Hong Kong	2016	Overloading	Roof collapsed

permanent loads from the intensive green roof system, such as substrate and foliage. It can also include hardscape such as benches and gazebos. Live loads include pedestrian load, storm water, wind load, equipment, temporary structures, etc.

3. **Plant selection.** Intensive green roof systems can provide the widest range of plant selection (including trees), due to the possibility of having a deep substrate depth. In comparison, extensive green roof systems have a more limited palette due to their shallow substrate, which may not provide sufficient anchorage for larger plants.

4. **Cost.** The cost of the green roof installation is a function of the system used as well as the installation area. When looking into the cost of each system, the Life Cycle Cost (LCC), comprising the installation, operation and maintenance, as well as disposal cost, should be considered for a more comprehensive assessment. Some indicative costs are shown in Table 9. Intensive systems may cost more due to the larger plants and furniture used. Cost should not supersede safety and maintainability concerns when deciding on the appropriate green roof system to use.

Table 9. Indicative Life Cycle Costs of green roof systems (Manso, Teotónio, Silva, & Cruz, 2021).

Cost	Extensive	Intensive
Installation ($€/m^2$)	67–128	156–627
Operation/Maintenance ($€/m^2$/year)	0.84–9.16	0.72–12.75
Disposal ($€/m^2$)	12	26

Extensive green roof systems

Extensive green roofs are characterised by their shallow substrate depth (less than 20 cm) and much lighter weight compared to intensive green roofs. This makes it a suitable alternative for older buildings or roofs that are not designed to support significant increases in loading. As a result of the thinner substrate, the plant palette will also be less diverse. Extensive green roofs tend to be non-accessible to the public and are not able to support pedestrian traffic.

The main components of extensive green roofs consist of the following (Fig. 34):

1. **Deck layer.** Strictly speaking, this layer is not part of the green roof system. However, it is the main component supporting the entire system, and prior assessment is required to ensure that the roof deck has the necessary insulation, waterproofing and protection to maintain the integrity of the roof during and after installation of the extensive green roof system.

2. **Drainage layer.** Due to the shallow depth of extensive green roof systems, excess water can be expelled easily. Some proprietary systems provide additional piping to ensure quick and efficient water drainage.

3. **Carrier layer.** This constitutes the bulk of the extensive green roof system: It provides a structure to house the irrigation, substrate, plants, etc. Some carriers are also designed to facilitate water drainage and storage via reservoirs. Carriers are modular and can take on a myriad of shapes and sizes. Some examples include egg-crate trays and roll-up mats.

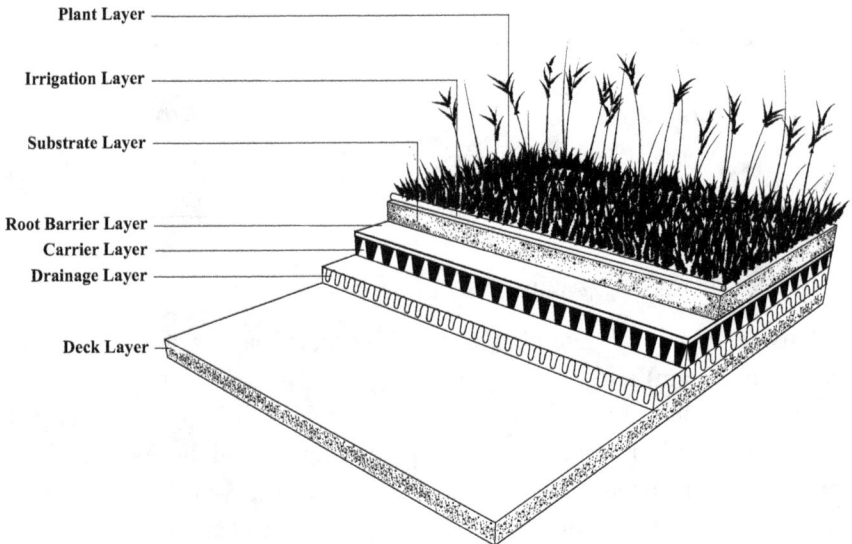

Fig. 34. Typical components of extensive green roof systems.

4. **Root barrier layer.** Similar to intensive green roof systems, the function of the root barrier layer is to prevent invasive roots as well as loose growing media from clogging the drainage and reservoir. Semipermeable geotextile filters are normally used for this layer.

5. **Substrate layer.** The exact constituents of a substrate layer for extensive green roof systems can vary tremendously. Since smaller plants are used for such systems, the substrate depth can be shallow and comprise materials that facilitate water retention and storage. Less top soil is used due to its high density. Instead, a mixture of peat, perlite, vermiculite or crushed brick may be used. Slow release fertilisers are normally added to ensure a steady amount of nutrients for plants over time. Some systems may be entirely soilless, consisting only of a mat as substrate (Fig. 35).

6. **Irrigation layer.** The design and installation of the irrigation for both intensive and extensive systems are similar, as it is a separate system that is normally not integrated with any of the other layers. Although the green roof may be naturally irrigated by rainfall, plant foliage may prevent it from reaching the substrate. Therefore, irrigation should still be provided.

7. **Plant layer.** The choice of plants for extensive green roof systems is more limited compared to intensive systems. Due to the shallow substrate, this system cannot support big shrubs or trees. The substrate composition tends to be light, porous, and, in general, much less compacted than typical soil mixes. This results in inadequate anchorage for plant roots. Common plants used for extensive systems include turf and ground covers such as *Cyanotis cristata*, *Sphagneticola trilobata* and *Dissotis rotundifolia*. Plant selection is also highly dependent on the system used: For mat systems without any substrate, only succulents and certain ground cover species can be used. Tray systems can hold more substrate and are able to support bigger plants such as *Schefflera arboricola* and *Excoecaria cochinchinensis*. Owing to the smaller plant palette, extensive systems may be less efficient at providing certain ecosystem services such as thermal regulation (less shade provision from plants) or stormwater retention (shallow substrate).

Fig. 35.   Gaia Mat™ extensive green roof system. (Images courtesy of GWS Living Art Pte Ltd.)

Some examples of proprietary extensive green roof systems can be found on www.ugl.sg/book/.

Extensive green roof systems are mostly proprietary, and the design of their components (e.g., carrier layers) may be patented. Detailed information on the systems can be obtained from vendors, who will normally provide both supply and installation services. This will include a selection of plants that are suitable for the particular system. Although it is not a requirement for designers to have detailed knowledge of the function and installation method of each component, some useful considerations for choosing the extensive system include the following:

1. **Function.** Extensive green roof systems serve very different functions compared to intensive systems. Owing to the shallow substrate depth as well as the thin carrier layer, most extensive systems are not able to support live loads for a prolonged period. Therefore, extensive green roof systems should not be used for roofs intended for social purposes or if heavy footfall is anticipated. Extensive green roof systems are suitable for non-accessible green roofs without any social function, where foliage can only be observed from a distance. They are also reasonably effective at reducing exposure to direct solar radiation and lowering thermal transmission into the building (Tan, Wong, Tan, Jusuf, & Chiam, 2015). However, the water retention capabilities of such systems will be much lower compared to intensive systems (MacIvor, Margolis, Puncher, & Matthews, 2013). For hybrid systems incorporating PV panels and greenery underneath, extensive green roof systems are often used as they are easier to install and take up less head room than intensive systems.

2. **Loading.** Extensive green roof systems are an ideal choice for retrofitting onto existing building roofs, or when the building roof is not designed to support significant additional loading. The lightweight setup of certain extensive systems allows flexibility for installation on sloped roofs as well as metal roofs. With thinner systems, more anchorage needs to be provided.

3. **Plant selection.** One limitation of extensive green roof systems is that plant selection may be rather limited. The shallow substrate depth does

not allow for large plants. Some systems can only accommodate succulents and ground covers, as they do not require a thick substrate to grow. To reduce the need for irrigation as well as maintenance, CAM plants such as succulents may be planted as their rate of growth is relatively slower, and they are more tolerant to drought conditions. Although these plants can still grow and provide good coverage over time, the green roof might look sparse as the leaves are small, and the plant height will remain low. This may adversely affect the aesthetic quality of the green roof. Should this be a concern, it is recommended to consider extensive green roof systems with thicker carrier layers and reservoirs so that it may accommodate a larger variety of plant species and sizes.

4. **Cost.** In general, extensive green roof systems are more cost-effective than intensive systems due to the thinner substrate layer, lower capital expenditure, irrigation and maintenance requirement (Table 9). This becomes progressively significant as the roof area increases (Berndtsson, 2010).

Plant

1. Benefits

Unlike green walls, where access to light and rain is highly dependent on the location of the wall, microclimatic conditions for green roofs tend to be quite consistent, with ample access to sunlight and rain. Therefore, plants are subjected to lower possibilities of light and water stress. The range of plants to consider for the green roof primarily relies on the type of system that is adopted for the project. Since light is typically not a limiting factor at the roof level, some provision for drought tolerance can be considered.

Another key consideration is to select plants that can provide specific benefits. If we examine the reason for the widespread adoption of green roofs in urban areas, it is mainly due to the increase in building regulations and urban development policies that aim to increase green cover. The purpose of which is to extract certain ecosystem services for the benefit of the urban population. These benefits can be for thermal regulation, improvement of the air and acoustic quality, water retention, preservation of biodiversity, etc. Details on common benefits can be found in the preceding section. The specific benefit to tap on is dependent on the type of plant used. For instance, shade provision and evaporative cooling are critical factors for temperature reduction. Selection for reducing temperature will therefore require an assessment of plant functional traits, such as the leaf size, canopy cover and plant evapotranspiration rate. If the objective is to promote biodiversity, fruit-bearing plants may be considered, to attract fauna.

Some attention may also be paid to the possible interaction between components which may adversely affect the overall quality of the green roof. For instance, if water retention is an expected benefit and the substrate is expected to be constantly wet, plants that are susceptible to root rot due to excessive moisture (e.g., *Peperomia nitida*) should be avoided. Similarly, brightly coloured plants such as *Guzmania bromeliad* may increase the aesthetic quality of the green roof, but water accumulated in the leaf axils are fertile breeding grounds for mosquitoes.

Therefore, when considering the types of plants to have on the green roof, it is important to identify the expected ecosystem service benefits and possible disservices that may arise.

2. Maintainability

Plant selection for maintainability is important for keeping the overall LCC of the green roof as low as possible and to ensure that benefits provided by the green roof will not be nullified by excessive costs or disservices associated with it. For instance, savings in cooling energy expenditure may be less than overall maintenance costs of the green roof, if plant replacement has to be done frequently. This defeats the purpose of having a green roof to reduce costs in energy consumption and is tantamount to greenwashing.

Another important aspect of maintainability is safety, which is a function of both planting design and plant selection. Plants on the green roof must be accessible in a safe manner, especially if they are located close to the roof edge. Trees of green roofs have to be set back from the roof edge depending on their height (i.e., the taller they are, the further they should be from the edge). Where possible, a maintenance work plan should be prepared to identify associated risks and mitigation measures.

Plants should be selected and placed on the green roof with the intention of minimising maintenance. This includes inspection, pruning and replacement. As green roofs may cover a large surface area, it is important to be able to visually assess the quality of plantings in a convenient manner. Therefore, taller plants can be avoided along the periphery of the green roof. The placement of plants should be done with consideration for easy access to facilitate pruning and replacement. This can be achieved by incorporating footpaths within the green roof so that all parts of the roof are accessible. Designers can also consider the interfacing between plants and hardscapes to minimise maintenance. For instance, instead of covering the entire green roof with turf, having a buffer (e.g., pebbles) between the turf and hardscape instead of letting the turf come in contact with them directly can significantly facilitate mowing.

The criteria for plant selection should include growth rate: Plants that grow too slowly will leave the substrate exposed and increase the chances of introducing weeds on the green roof. Plants that grow too quickly will require more frequent pruning and may become invasive if left unattended, encroaching onto other plants and threatening their survival. To minimise the impact of water stress of plant replacement, plants that require less water, such as succulents, can be considered.

For hybrid systems such as PV and green roof setups, plant selection is critical to avoid any adverse impact on the efficiency and safety of PV panels. Since the headroom under the PV frames may be limited, plants selected for such systems cannot be too tall. Climber plants should be avoided, as they may climb up the PV panels over time and over the PV cells.

An alternate approach to urban landscaping is to introduce spontaneous vegetation, where the eventual plant palette includes other species that naturally grow on the roof garden and are not removed. Their seeds may be carried by the wind or other animals. They are left to grow and establish themselves naturally on the green roof without any intervention. Maintenance is kept to a minimum (CUGE, 2015). This method provides a more natural aesthetic to the green roof and requires lesser upkeep. Due to the wider variety of plants available, spontaneous vegetation can also help promote biodiversity more effectively compared to conventional mono-culture green roofs. Note that green spaces incorporating spontaneous vegetation tend to look unkempt in appearance due to the deliberate avoidance of maintenance (pruning, weeding, etc.).

Chapter 5

Plant-scale greenery

This chapter focuses on providing practical information on how to acquire and interpret data for implementing greenery in and around buildings. At this scale, it is important to consider the local microclimate, characteristics of plants species, as well as the consequences of their interaction. Basic requirements of plants for survival are first introduced, followed by a discussion on potential limiting factors in the urban environment. The subsequent section touches on various methods of measurement, with the aim of providing an accurate, quantitative description of the environment before deciding on the appropriate landscape design or planting strategy. Reference profiles of common abiotic parameters as well as a flow chart, are provided to aid users in formulating a robust plan for data acquisition and evaluation to support design decisions.

Some key questions for this chapter include:

1. What do plants need to survive?
2. What does the environment provide?
3. How do we know if they are sufficient?

Physiological requirements of plants

For greenery to be effectively situated in the built environment, it is necessary to understand what plants need to survive. Unlike edible crops, where yield optimisation is imperative, ornamental plants are often planted without the need for achieving fast growth. In fact, faster growth may lead to higher maintenance requirements, which will increase the overall Life Cycle Cost (LCC). Nevertheless, it is important to acquire a basic understanding of plant physiology for proper deployment.

What is a plant? Broadly speaking, a plant is a living organism made of cells organised into tissues, forming organs such as leaves, flowers, stems and roots. To survive, plants make food via the photosynthetic process, where energy from the sun is absorbed through their leaves, converted into sugar and transported to other parts of the plant (Fig. 36).

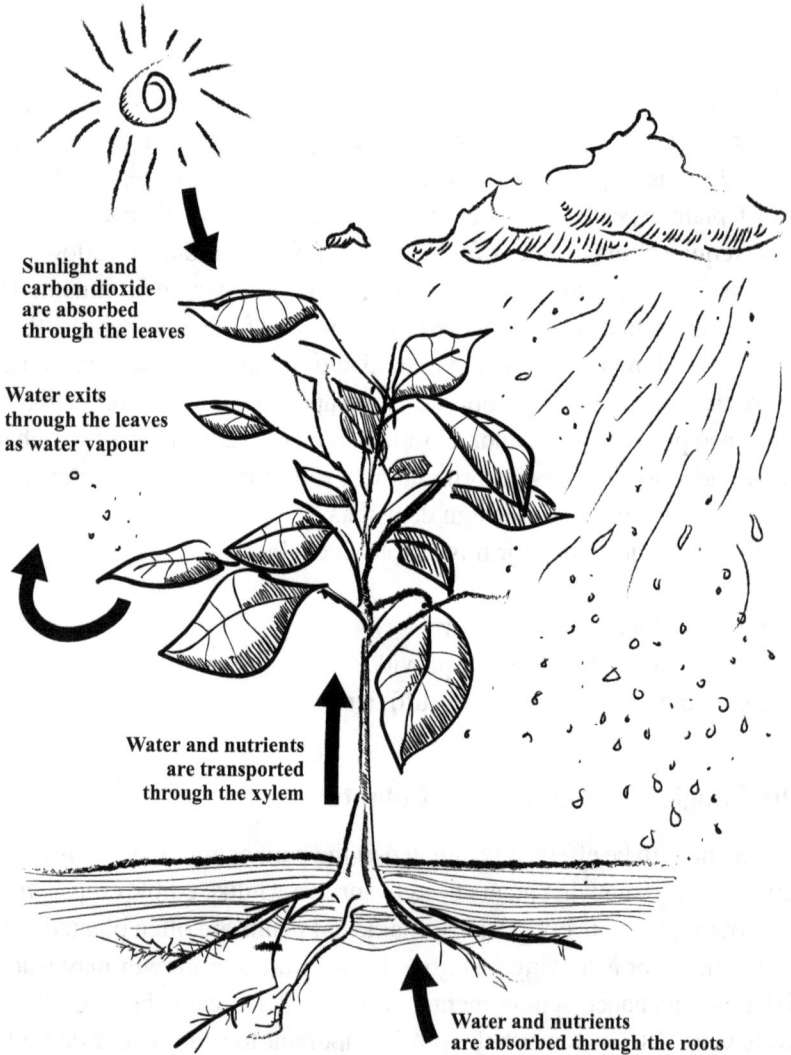

Sunlight and carbon dioxide are absorbed through the leaves

Water exits through the leaves as water vapour

Water and nutrients are transported through the xylem

Water and nutrients are absorbed through the roots

Fig. 36.   Photosynthesis process, showing the movement of water and nutrients throughout the plant.

Here, we look at the main components that govern food production in plants and how they can be affected by the urban environment. To make food, plants have to undergo photosynthesis. The simplified equation for photosynthesis can be described as follows:

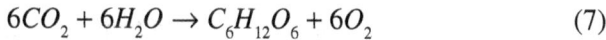

$$6CO_2 + 6H_2O \rightarrow C_6H_{12}O_6 + 6O_2 \tag{7}$$

or rephrased as:

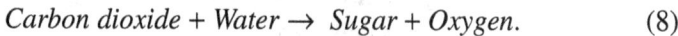

$$Carbon\ dioxide + Water \rightarrow Sugar + Oxygen. \tag{8}$$

In addition to carbon dioxide and water, light energy is required to provide energy to trigger this reaction. Ambient conditions must be suitable for photosynthesis to occur (e.g., not being too hot and having sufficient sunlight). While these components are limiting factors for plant growth, some may not be of much concern. For instance, carbon dioxide is typically abundant in the atmosphere and is not a limiting factor for plant growth, nor is it adversely affected by urban conditions. The supply of water, on the other hand, is highly reliant on its source, and availability may be subject to changing weather conditions. Therefore, these components need not only be present but also assessed in tandem with their surrounding environment.

Light

Energy required for photosynthesis is primarily provided by the sun. In the urban environment, light distribution is heterogeneous and subject to self-shading and overshadowing from adjacent built structures. Excessive irradiance may also result in photoinhibition as well as cellular damage in plants. This heterogeneity is the core limiting factor in the plant selection and placement process.

Water

Water is an essential component for many biochemical reactions that occur during photosynthesis. Water is absorbed through the roots of the plant and travels through its xylem to the leaves of the plant. Water exits the plant through its leaves as vapour.

In practice, water may be a limiting factor when:

1. Plants undergo extreme drought events and do not receive manual irrigation. This may happen even in tropical zones with consistently high annual rainfall.
2. There is irrigation system failure, resulting in water not reaching the plants. Failure may stem from any part of the irrigation system, such as pipes, pumps, water tanks, etc.
3. Substrate is compacted, which prevents water from reaching the roots. This occurs naturally with time, especially for top soil.

Mineral nutrients

Nutrients are required to facilitate plant growth and survival. Conventionally, they are broken down into macronutrients and micronutrients.

Macronutrients: Required in large quantities. Examples include nitrogen (N), phosphorus (P), potassium (K), calcium (Ca), magnesium (Mg) and sulphur (S).

Micronutrients: Required in trace amounts. Examples include boron (B), copper (Cu), iron (Fe), chloride (Cl), manganese (Mn), molybdenum (Mo) and zinc (Zn).

In natural settings, these elements can be found in soil or organic matter. In urbanised environments where plants may be confined in planter boxes, there may be deficiency in some nutrients. This may be addressed by adding fertilisers.

Describing the urban environment

Why measure

In addition to plant physiology, it is also necessary to assess plant selection and placement in tandem with its intended mode of deployment. In the previous chapter, we looked at how greenery interfaces with the ground, along waterways and on building surfaces. These can be

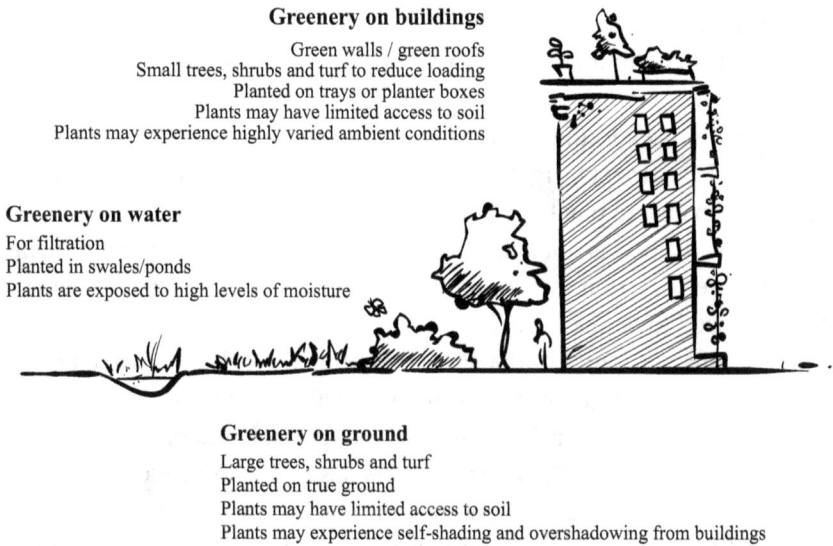

Greenery on buildings
Green walls / green roofs
Small trees, shrubs and turf to reduce loading
Planted on trays or planter boxes
Plants may have limited access to soil
Plants may experience highly varied ambient conditions

Greenery on water
For filtration
Planted in swales/ponds
Plants are exposed to high levels of moisture

Greenery on ground
Large trees, shrubs and turf
Planted on true ground
Plants may have limited access to soil
Plants may experience self-shading and overshadowing from buildings

Fig. 37. Interfacing between greenery and different urban components, showing the influence of plant selection and physiology.

considered as system-based considerations. The plant palette and installation method is dependant on its reciprocal urban component (Fig. 37). Ground level greenery, for instance, can accommodate all forms of greenery, whereas trees cannot be planted on some extensive green roofs due to the shallow substrate depth. Greenery on buildings will tend to have uneven diurnal exposure to solar irradiance due to self-shading or overshadowing. These system-based considerations can dictate plant selection and will also influence plant physiology.

It is, therefore, important to be able to describe the urban environment with some level of accuracy before plant selection and placement is finalised. In doing so, we will be able to assess prevailing microclimatic conditions and specify plants that grow well under these conditions. Less maintenance (plant replacement) will be required.

What to measure

On the specific parameters of the urban environment to quantify, two factors are considered:

1. What plants need to survive.
2. What the environment can provide for plants to survive.

Going to back to the equation for photosynthesis (Equation (7)), carbon dioxide, nutrients, light, water and suitable ambient conditions are necessary to ensure plant growth.

Carbon dioxide is not a limiting factor as it is present in the air. Plants do not suffer due to a lack of carbon dioxide. Therefore, this parameter does not need to be measured.

Nutrients are found in the soil or substrate, which are containerised. When designing for greenery, designers will also specify the container for housing plants. This can be planter boxes, pots of indoor greenery and trays for roof gardens. The onus is on the designer and vendor to provide soil or substrate suitable for plants. This is not a parameter that is normally measured at the design stage. In the installation and maintenance stage, soil compaction, as well as pH level, may influence plant health and should be monitored where possible.

Lighting exposure varies significantly in the urban environment, firstly due to under-exposure from self-shading and overshadowing from built structures, as well as over-exposure from high placement. This variability will drastically influence the photosynthesis process and the ability of plants to make food. Where possible, this parameter should be measured.

The availability of water at a given site is highly contextualised. Areas open to the sky may be naturally irrigated by rainfall, which may be sufficient for much of the year. Fully or partially sheltered areas will require manual or automatic irrigation set up to ensure sufficient water for plants. Measurement of water content can provide information on the availability of water in the surveyed area.

Ambient conditions include vapour pressure deficit, wind speed, surface (if plants are in contact with building surfaces) and mean radiant temperature. These variables have to be within their respective acceptable ranges for plants to thrive, but since the focus of this book is on ornamental plants and plant yield is not a criterion, a larger tolerance is acceptable. These variables are not as critical as light and water. Measurement can be considered if extreme microclimatic conditions are expected for the planting area, such as rooftop gardens with high exposure to direct sunlight or plants are in direct contact with metal surfaces (e.g., façades).

How to measure

There are two modes of data acquisition: field measurement and simulation.

Field measurement

Field measurement can be conducted when:

1. You have access to the site

Some areas, such as rooftops, may be inaccessible to the public and require approval prior to the visit. Some areas may be out of reach. For instance, it may not be feasible to measure anything from the top of a green wall that is 20 m high.

2. You have the appropriate equipment for measurement

Different sensors are needed to measure different variables. It is important to identify the specific variable that you wish to measure (e.g., air temperature) and the resolution that is required for the objectives of the measurement.

3. You have permission to install equipment, or to perform measurements onsite

Permission has to be sought from facility managers or building owners prior to entering the site, as well as to install any equipment. There may be multiple stakeholders who are not aware of your intention to survey the premise.

4. You are confident that installed equipment will not be stolen or damaged

Survey equipment may be costly and difficult to replace if stolen or damaged. The level of security around the area of interest should be assessed before deciding on whether sensors can be installed without fear of tampering. Some common methods of minimising the risk to equipment loss are to install sensors above 2 m, to pick measurement spots that are inaccessible to the public and to include warnings or dummy CCTV cameras to deter tampering.

Light

In-situ light measurement is a reliable method for ascertaining light provision for an area and the corresponding plant palette. Light levels should be measured before planting. There are two aspects of light exposure that need to be measured: intensity and duration.

Intensity

Before quantifying light intensity, the appropriate metric needs to be selected. Not all of the sunlight absorbed by the plant can be used for photosynthesis. Plants only absorb light from the 400 nm to 700 nm wavelength, which is a subset of the electromagnetic spectrum. This range is termed Photosynthetically Active Radiation (PAR) (McCree, 1981). This is a suitable metric to use when deciding whether a specific location has sufficient lighting.

Common sensors used for measuring lighting intensity for plants

PAR sensor: PAR sensors measure PAR directly. They are the best option for assessing light provision for plants.

Spectrometer: Spectrometers provide spectral measurements such as Photosynthetic Photon Flux Density (PPFD) and irradiance (W/m²). Detailed spectrum charts may also be available. These sensors are typically handheld and used for spot measurements. Readings for PAR may be obtained directly from such sensors.

Lux meter: For plants to thrive, PAR should be the metric used to assess light levels. In practice, this is often not the case, as there are more common and cheaper sensors for measuring lighting levels. Lux level, for instance, is more commonly used to determine the brightness (or illuminance) of an area. It is also possible to download free lux meter apps to use on mobile phones. Lux level may be highly correlated to the solar irradiance level and PAR as they are all indicators of brightness, but it should not be used for assessing plant lighting availability. There are potentially severe implications for using lux level for estimating lighting levels for plants:

1. Over-estimation of lighting — Lux level is a quantitative measurement of light intensity perceived by the human eye. It does not directly describe the amount of PAR that is available. An area that is perceived to be well illuminated may have high lux levels but low PAR levels (Runkle & Bugbee, 2013). This is a common occurrence in semi-outdoor or indoor settings, where artificial lighting may be present.

2. Over-compensation of supplementary lighting — When designing for supplementary lighting for plants (especially indoor plants), lighting designers may adopt a generalised approach of providing at least 1,000 lux for the area. A value of 1,000 lux is used because it is deemed equivalent to lighting levels observed on a day with overcast sky conditions, following the logic that since plants have been observed to survive under such conditions in outdoor settings, they should also be able to survive with artificial lighting. However, a large number of growth lights may be required to achieve an average value of 1,000 lux across the entire green area, which will lead to excessive energy consumption. It is, therefore, necessary to convert lux readings (photometric) to PAR (radiometric) for the evaluation of lighting availability for plants.

   Tan and Ismail (2016) proposed the conversion from lux to PAR in a tropical setting (Singapore):

$$PAR = 0.021 * L, \qquad (9)$$

where

PAR = Photosynthetically Active Radiation (mol/m^2/d)
L = Lux (Mlux/d)

Duration

Consider an area (A) that receives intense daylight for 1 hour a day and another area (B) that receives half the intensity but for 4 hours a day. Which area has better overall lighting conditions for plants?

The above scenario demonstrates the importance of duration in determining overall light provision for plants. In addition to light intensity, it is equally important to consider the duration of exposure. The recommended method of doing so is to calculate the Daily Light Integral (DLI), which describes the amount of photosynthetically active photons that are delivered to an area over a 24-hour period.

The DLI for an area can be calculated using the formula below (Tan, Wong, Tan, Ismail, & Wee, 2017):

$$DLI = PAR * 0.0036 * T, \qquad (10)$$

where

DLI = Daily Light Integral (mol/m^2/d)
PAR = Photosynthetically Active Radiation (μmol/m^2/s)
T = Duration of exposure (Hr)

In practice, the calculation of the DLI is often overlooked because surveyors tend to only take spot measurements onsite, which only provides information at a particular point in time.

Some indicative ranges are shown below:

- 5 to 50 mol/m^2/d in temperate regions (Korczynski, Logan, & Faust, 2002)
- 1 to 35 mol/m^2/d in greenhouses (Dorais, 2003)
- 2 to 35 mol/m^2/d on building façades during clear sky conditions (Song, Tan, & Tan, 2018)
- 29 mol/m^2/d in a tropical urban setting (Tan & Ismail, 2014)

There are two options for carrying out plant selection after obtaining the DLI of the site. The first option is to select plant species according to their DLI requirements. For instance, if the DLI of the site is 4 mol/m^2/d, it will be suitable for *Philodendron erubescens* (Tan et al., 2017), which requires 1 mol/m^2/d. It will not be suitable for *Ocimum basilicum L.*

Table 10. Plant lighting requirement (Torres & Lopez, 2010).

Lighting requirements	DLI range (mol/m²/d)
Full sun	>12.0
Partial shade	>6.0–12.0
Full shade	3.0–6.0

(sweet basil) (Chang, Alderson, & Wright, 2008), which requires at least 10 mol/m²/d. The limitation of this method is that an extensive catalogue of plant DLI data needs to be available. Currently, there is a paucity of information on lighting requirements for ornamental plants commonly used for landscaping. Most related data tend to be for edible crops, where lighting significantly affects yield.

The second option is to refer to a generalised categorisation for selection, such as in Table 10. Instead of knowing the specific DLI of the species, plants are grouped into full sun, partial shade and full shade plants. This method relies more on empirical observation over time, but the knowledge base is large enough to be used with confidence in practical guides (Boo, Omar-Hor, Ou-Yang, & Ng, 2003).

Water

The availability of water directly impacts the photosynthesis process. Ensuring that plants are well-irrigated is therefore required to ensure plant health. The impact of heat stress on plants can be quite obvious. Some common examples include flaccid stems or discoloured leaves. Visual inspection, while convenient, may result in irreparable plant cellular damage if not caught in time.

The approach to monitoring water availability is different from that of light provision. For light, an assessment should be conducted prior to planting. We assume that lighting levels will not deviate significantly from initial measurements (assuming the measurement method is robust). For water, regular monitoring is recommended after planting is complete. This is because water availability can fluctuate wildly, depending on prevailing weather conditions: one moment, the substrate can be saturated with water after rain, and the next moment it can be completely dry during a drought period or if the irrigation system is faulty. In tropical climate, faulty irrigation may result in severe plant health degradation in a very short amount of time. Without constant monitoring and timely intervention, plant death and replacement is unavoidable.

When we try to measure water availability, we are really quantifying the amount of water in the soil or substrate that acts as the base layer where plant roots are anchored. This is commonly known as soil moisture. It is where water and nutrients are absorbed by the plant to facilitate photosynthesis (Fig. 36). It is not necessary to estimate the amount of moisture in the air (relative humidity).

Common sensors used for measuring moisture content in substrates

Broadly speaking, the amount of water available in the substrate can be expressed in terms of its volumetric water content (Kirkham, 2014). This describes the volume of water present in a given sample.

$$\theta = \frac{V_w}{V_t}, \tag{11}$$

where

θ = Volumetric Water Content (m^3/m^3)
V_w = Volume of water (m^3)
V_t = Volume of sample (m^3)

Based on Equation (11), the theoretical maximum value will be 1, or 100% of water by volume within the same volume of sample. Practically, values of 0.5 m^3/m^3 or 50% are common. Indoor plants that are not exposed to direct solar radiation or fluctuating Vapour Pressure Deficits (VPDs) need at least 0.1 m^3/m^3 of volumetric water content to be free of water stress. Note that the objective here is to provide sufficient water for plants to survive and not to accelerate growth.

The appropriate sensor to use is largely dependent on the objective of the measurement. For scientific studies or studies requiring high precision, dielectric-based sensors are recommended. For monitoring in a facility management capacity, resistance-based sensors may be sufficient.

There are a few methods of sensing soil moisture. The first method is to periodically survey the area of interest with a soil moisture probe and obtain spot measurements or download data if the sensor is deployed with a data logger. This method is labour-intensive but good for checking all related building services. The second method is to connect the soil moisture sensor to a Building Management System (BMS) or dedicated server so that soil moisture content is continuously monitored and can be tracked online. An alarm can also be set to be triggered if the volumetric water content falls below a certain threshold or if the rate of decrease is faster than normal. This could then activate an automated irrigation system to increase soil moisture levels to a desired level. This method is more effective at addressing issues of plant stress due to water but requires higher investment in equipment, a reliable power source and an internet connection. At the time of writing, there are many products available offering soil moisture measurement with Internet-of-Things (IoT) functionality.

A common issue faced when assessing moisture content of green walls is that higher parts of the walls are less accessible. It may be physically inconvenient or even dangerous to install sensors at these areas. In such cases, infrared thermal imagers may be used to quickly assess the moisture content of the wall. Due to the presence of water, most of the green wall surfaces will be cooler. Dryer surfaces will tend to have a slightly higher surface temperature. The difference can be seen clearly in a thermal image (Fig. 38).

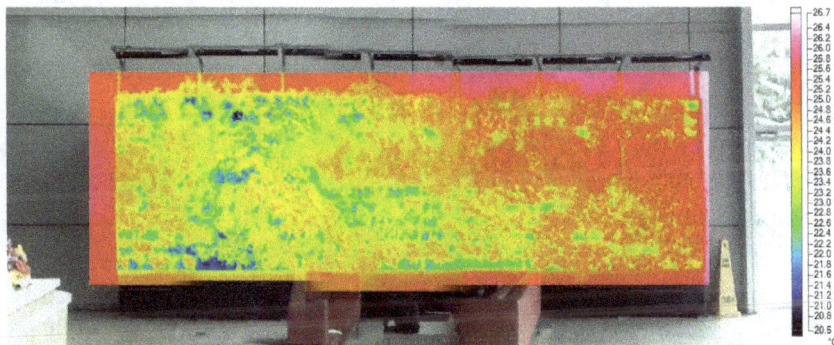

Fig. 38.   Infrared thermal image, showing the dryer part of a green wall (top right) in red.

Placement

Soil moisture sensors should be installed in locations that can provide the most accurate information of water content for the planted area. Water distribution in soil or green walls should not be assumed to be homogenous, especially when plantings do not have or only have partial access to rainfall and rely on automated irrigation. It is common to find planter boxes that are saturated with water on one end and completely dry on the other. For green walls, irrigation pipes are installed intermittently, with water trickling down from emitters and reaching the lower parts of the wall via gravity. In practice, it is common for areas further away from the emitters to receive significantly less irrigation.

Where possible, sensors should not be placed too close to irrigation drip emitters: areas close to drip lines tend to be perpetually moist simply due to their proximity to the water source. They should be placed at least a few inches away from the emitters (Fig. 39). Multiple sensors should also be deployed throughout the area of interest for more accurate results.

Plant health

Measurements of plant physiology offer a direct way of tracking its health. It eliminates possible errors in judgement through indirect abiotic measurements. While plant physiology may seem to be an obvious parameter to monitor, some distinctions are to be made between plants that are grown to be harvested (crops) and ornamental plants that are used for landscaping the built environment: More attention is paid to maximising yield for the former, while the latter can tolerate a larger range of conditions and stress, as long as they do not experience permanent cellular damage.

The following points should be considered before embarking on plant health measurements:

1. Normally, plant health (or stress) is measured in a predominantly homogenous setting with low species variation, and results will be representative of a large area, such as a farm. This implies that the health of the surveyed plant is indicative of the state of a larger sample of plants. For instance, if measurements indicate that a plant

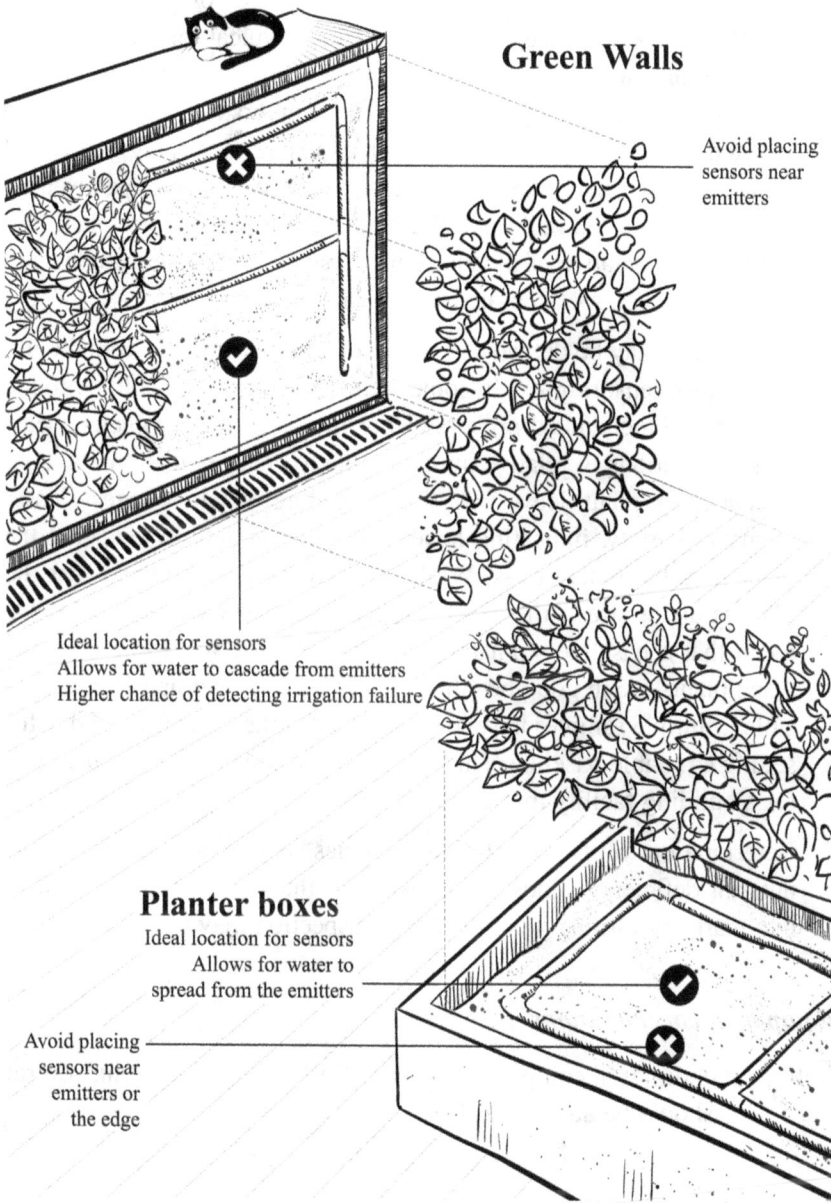

Green Walls

Avoid placing
sensors near
emitters

Ideal location for sensors
Allows for water to cascade from emitters
Higher chance of detecting irrigation failure

Planter boxes

Ideal location for sensors
Allows for water to
spread from the emitters

Avoid placing
sensors near
emitters or
the edge

Fig. 39.   Considerations for soil moisture sensor placement.

is under stress, it would mean that the larger area of plants may also be affected (perhaps the entire farm). This is typical for crops or high-value ornamental plants in a farm setting. In the urban environment, conditions can be highly heterogeneous: plant boxes can be small and situated throughout the building, with exposure to varying degrees of light and rain. There may also be a large variety of plant species used for landscaping the building. The health conditions of one plant will not be representative of the entire plant palette. Multiple samples will have to be taken for accurate assessment.

2. Plant stress may be detected, but the source of stress may remain elusive. Most sensors provide some indication of plant health through measuring biochemical processes. For instance, the porometer measures fluctuations in relative humidity due to changes in stomatal conductance in plant leaves. A lower reading would suggest lesser stomatal activity, indicating lower transpiration activity and that the plant may be under stress. While we may have ascertained that the plant is under stress, the exact cause may not be clear. It would also mean that a baseline stomatal conductance value has to be obtained when plants are in good health so that meaningful comparison is possible. Multiple samples have to be obtained at the same time of day for setting a reliable baseline. This is manpower-intensive and impractical for a building facility management outfit.

3. Deterioration to plant health due to issues such as pest infestation or nutrient deficiency tends to be visible to the naked eye (leaves with bite marks or turning yellow). Visual inspection may be sufficient.

Equipment considerations

Before embarking on equipment selection for field measurement, several key issues should be addressed (Fig. 40):

Biotic or abiotic measurements

Are you measuring plant or environmental variables? The appropriate equipment should be used for the parameter of interest. Biotic sensors will require direct contact with the plants, while abiotic sensors may be installed some distance away.

Abiotic
measurements

Supporting structure must be provided
Mounted on tripods or existing structures

Outdoor
measurements

Sensors and loggers have to be weather-proof

Continuous
measurements

Provides more comprehensive data
For setting diurnal or monthly profiles
Data needs to be downloaded periodically
Sensors need to be checked regularly
Static measurements

Biotic
measurements

Needs to be in direct contact with plant
Multiple samples required
Baseline readings required

- Mounted on existing structure
  Raised to avoid tempering

- Regular inspection
  is recommended

- Multiple sensors
  can be linked to
  a single logger

- Sensor

Logger should •
be locked

- Signage with
  contact
  information
  should be
  provided

Instantaneous
measurements

Convenient but only
provides information
for that instant
Mobile measurements

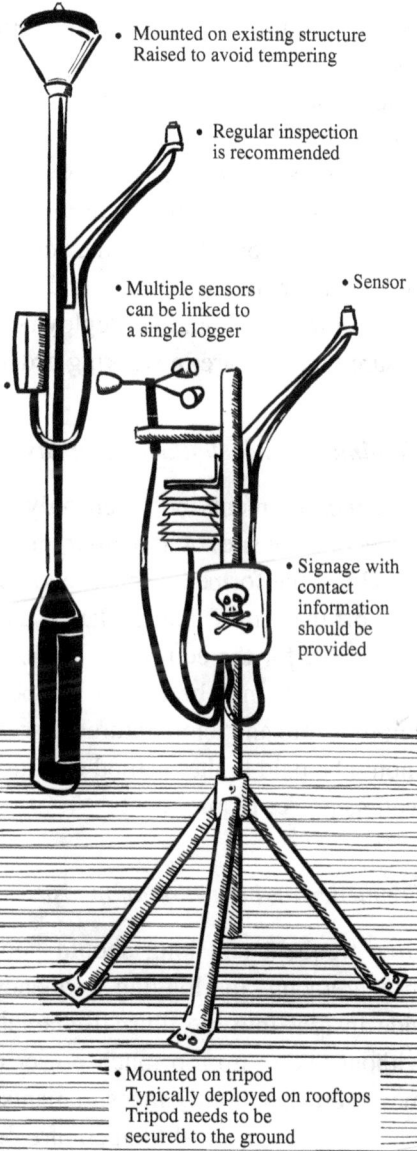

- Mounted on tripod
  Typically deployed on rooftops
  Tripod needs to be
  secured to the ground

Fig. 40.   Equipment considerations.

Indoor or outdoor measurements

Will the equipment be located indoors or outdoors? Most equipment will specify the recommended mode of usage. Some sensors designed for indoor use may be deployed in outdoor conditions, given proper housing. For instance, indoor temperature sensors can be deployed outdoors when placed in a protective solar shield. Outdoor sensors are subjected to weathering and external interference (e.g., vandalism). Therefore, it is important to consider how the sensors will be installed, how they can be accessed for data download and maintenance, and how they respond to degradation from outdoor deployment. Regular inspection is necessary to ensure that they are in working order.

Continuous or instantaneous measurements

Will measurements be taken only at an instance (e.g., 1 reading when inspecting the site) or at regular intervals (e.g., every hour, 24 readings per day)? For the former, a sensor would be sufficient. Measurements can be read off the sensor. For the latter, a logger is required to store the data. It is possible for the equipment to have both sensing and logging capabilities. Loggers may be able to support multiple sensor inputs (e.g., light, soil moisture and temperature sensors all plugged into 1 logger). Attention should be paid to the method of data acquisition, data storage capacity, as well as power requirements.

Frequency and duration of measurement

How frequent and long should your measurement be? For continuous measurements, the logging frequency should be sufficiently high to provide insight into the diurnal characteristic of the parameter in question without taking up too much power or storage space. For environmental measurements, hourly measurements tend to be sufficient for establishing daily profiles. Typical temperature or light sensors with logging function can operate for months under normal conditions.

The duration of measurement is dependent on the variable of interest. For instance, lighting measurements conducted for the purpose of checking light provision prior to planting may require a few months' worth of measurement at hourly intervals. Soil moisture needs to be continuously measured to ensure no disruption in plant irrigation.

Simulation

The term "simulation" refers to the imitation of certain real-world processes via models that represent key characteristics of these processes (Banks, 1998). Simulation studies can be considered when:

1. Access to site is not possible

For projects still under development, simulation is the only viable way of estimating variables of interest.

2. An extensive dataset is required

Environment parameters such as light levels are highly contextualised and do not stay constant throughout the year. Therefore, a dataset spanning at least one year should preferably be used for the analysis of light provision. Due to time and resource constraints, it may not be possible to measure lighting conditions onsite for 12 months. This limitation is easily addressed by most simulation software, which offers the option to customise the period of simulation. Climatic data can be obtained via online weather data repositories.

3. There are constraints to equipment deployment

For sites that are too big, it may not be possible to deploy sufficient sensors to obtain good-quality data for analysis (e.g., for a township). Safety concerns may also hinder deployment, such as large green walls spanning several storeys above ground. In such cases, simulation is a safer and more convenient alternative to field measurement. For most simulation software, there is no limit to model size and sensor quantity.

4. Impact of greenery needs to be estimated

In addition to simulating urban conditions for maximising plant health, it is possible to simulate the numerous benefits of greenery. Ecosystem services provided by greenery, such as temperature reduction, can significantly improve outdoor thermal comfort in the built environment. It is possible to simulate the impact of temperature due to the presence of greenery. By understanding the amount of cooling that can be achieved, designers can improve their designs such that cooling due to greenery can be maximised.

Pros and cons of adopting a simulation approach

There are several advantages to opting for simulation studies instead of field measurements. Firstly, simulations can be processed quickly and distributed over multiple processors. For fieldwork, the maximum amount of data that can be acquired is dependent on the measurement period: one month of measurement will get you one month of data (or less). With simulation, a year's worth of solar exposure can be simulated in hours. This makes it possible to conduct iterative studies such that multiple design options can be evaluated for an optimal solution to be derived. This is termed "Generative Design", where 3D models are created and optimised by computer software.

Secondly, multiple sources of simulation results can be assessed in tandem. For instance, lighting simulation results can be evaluated together with temperature or wind simulation results to understand the impact of the design on multiple factors such as light provision, cooling potential, ventilation, etc. This enables a more comprehensive approach to evaluating your design.

Equipment required for simulation is a lot lesser compared to field measurement. Only the computer and relevant software are required for simulation work, whereas for fieldwork, the right equipment has to be procured, set up and installed for data acquisition.

Although simulation studies offer a quick and convenient way of sensing the urban environment, some limitations have to be acknowledged. Firstly, while actual simulation can be fast, pre-simulation preparatory work can be tedious and time-consuming. The area of interest has to be modelled with sufficient detail, and accurate simulation inputs need to be available. These requirements are highly specialised and may require expert input. Simulation models tend to be overtly simplified: Buildings are often modelled as rectilinear blocks without any façade detail. Trees and shrubs are represented as simple geometries without branch or leaf detail. Plant functional traits such as the Leaf Area Index or transpiration rates are approximated from the literature (Gromke, Blocken, Janssen, Merema, van Hooff, & Timmermans, 2015; Ouyang, Morakinyo, Ren, & Ng, 2020). Such simplification and approximation are often necessary to reduce computational load, but this will inevitably lead to inaccuracies in appraising the simulated environment.

What to simulate

The first category of simulation involves abiotic parameters and quantifying their impact on plant health. Similar to fieldwork, not all environmental parameters can or need to be simulated. Nutrient levels in the soil and CO_2 levels in the air need not be simulated as both tend to be in abundance. Irrigation is typically provided for areas with no access to rainfall, so water availability does not need to be simulated. Light provision, on the other hand, will vary according to location and adjacent structures. Light exposure is, therefore, a variable that can be simulated, with its results used for evaluating light provision for plants.

The next category of simulation involves quantifying the impact of plants on the urban environment. For instance, simulation might be conducted to estimate the impact of trees on the surface temperature reduction of pavements in a park or how placement of trees and shrubs can influence ventilation in the surrounding area (Buccolieri, Santiago, Rivas, & Sanchez, 2018). For this category, the more commonly simulated parameters are the mean radiant temperature, surface temperature, air temperature and wind (Kong et al., 2022; Salata, Golasi, de Lieto Vollaro, & de Lieto Vollaro, 2016; Wong et al., 2007). Such simulations are useful for evaluating design proposals aimed at improving thermal comfort conditions in the urban environment.

How to simulate

The recommended workflow for conducting simulation is shown in Fig. 41. Firstly, a virtual model of the area of interest has to be created. Commonly known as a 3D model, it can done using modelling software such as AutoCAD, Rhinoceros 3D, ENVI-met, Sketch-Up, etc. These models provide a virtual representation of objects in the physical world.

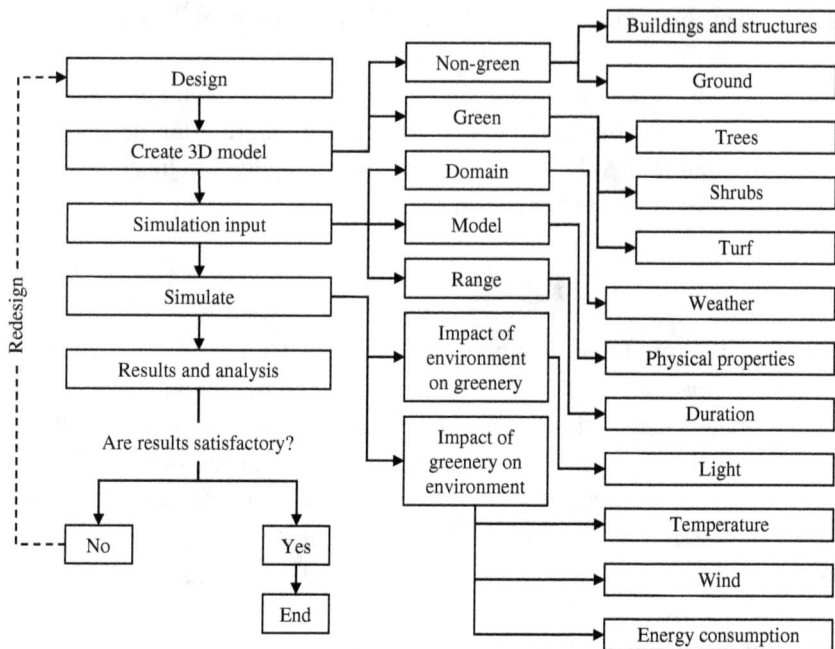

Level of Detail (LOD)

LOD 1
Conceptual

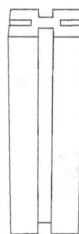
LOD 2
Approximate geometry

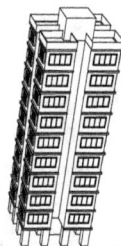
LOD 3
Precise geometry

Trees and shrubs

Point

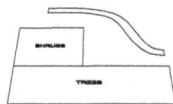
2-dimensional

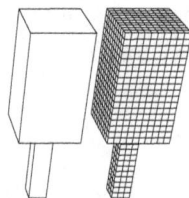
3-dimensional

Fig. 41. Workflow for simulation, LOD and different ways of modelling trees and shrubs.

Some commonly modelled items include buildings, shelters, roads, furniture, trees and shrubs. In principal, only objects that will significantly influence the simulation need to be modelled.

The modelling boundary has to be established. This boundary denotes the range to which models will have influence on dependent simulation variables. For instance, a simulation of indoor light provision for a single room does not require modelling the entire building — a room would suffice. For the simulation of light provision of a green wall, the building, as well as adjacent buildings, may need to be modelled as self-shading, and overshadowing can occur.

There are many ways to model objects. Some examples include Computer-Aided Design (CAD) modelling, Building Information Modelling (BIM) and Geographic Information Systems (GIS) modelling. In CAD modelling, 3D models are created by constructing points, lines and surfaces on a computer platform. There is no formal differentiation between the objects (i.e., a box represents a box, the other box also represents a box). In BIM, the same objects can have specific properties assigned to them (i.e., this box represents a door, the other box is a table). In GIS modelling, models can have geospatial data assigned to them (i.e., these dots represent trees and data on species, girth and height, and the date of planting is included). The Level of Detail (LOD) required for non-greenery objects varies according to the scale and objectives of the study (Biljecki, Ledoux, & Stoter, 2016) (Fig. 41). Generally, the level of detail for a model decreases as the scale of modelling increases. For city-scale modelling, buildings can be represented as rectilinear blocks with no façade details (Yang, Zhang, Ma, Xie, & Liu, 2011). For building scale modelling, details such as overhangs, parapets and cantilevers may need to be modelled as they can significantly influence simulation at this scale (Lee et al., 2013). For indoor lighting, furniture, sunshades and blinds may need to be modelled as they will influence overall lighting quality (Bustamante, Uribe, Vera, & Molina, 2017). Similarly, the LOD of greenery objects differs according to the model scale. In addition to scale, different forms of greenery are modelled differently, depending on the simulation purpose. At the city scale, trees may only be represented as points, whereas for wind simulation, the specific tree canopy shape needs to be modelled (Kang, Kim, & Choi, 2020).

After the modelling is complete, relevant simulation inputs will have to be appended. These inputs are required for both the modelled objects as well as the simulation domain.

For modelled objects, physical properties such as thermal storage, opacity and reflectivity can be assigned to relevant parts of the model (Harish & Kumar, 2016). These properties will enable accurate interaction between simulation parameters. For instance, if the building surface reflectivity is set to high, it will reflect more light from its surroundings and provide a more accurate estimation of the overall lighting quality of its surroundings. For greenery-based objects, specific plant functional traits such as the plant evapotranspiration rate and canopy albedo may be assigned. In this manner, the physical properties of greenery and non-greenery objects can be differentiated: both a normal wall and a green wall can be modelled in the same way but can be assigned different input values so that their simulated impact will be reflective of their respective physical properties.

For the simulation domain, it is necessary to set the general conditions for the simulation environment. The most common domain to set is the weather condition, which is based on the simulation location. The weather file will include data such as diurnal temperature profiles, solar irradiance data, etc. (Herrera et al., 2017). This is an important aspect of simulation, as factors such as light provision are highly affected by the sun's path. Typically, the appropriate weather file has to be selected and the duration of the simulated specified before simulation can begin. For Computational Fluid Dynamics (CFD) simulation, additional inputs such as wind speed and direction have to be provided to ensure the accuracy of results (Kang, Kim, & Choi, 2020; Zheng et al., 2020).

Simulating the impact of the environment on plants

A common objective of conducting simulations is to estimate the impact of the environment on plant health. As previously discussed, it is important to first identify what plants need to grow well and whether these variables are affected by changes in the environment. Similar to field measurement, light provision is the main variable that is simulated. Water, CO_2 and nutrient availability are normally not simulated.

Most simulation involving light is conducted by considering the impact of solar irradiance exposure. By simulating the amount of solar irradiance incident on a surface, information on lighting level, as well as heat absorption can be calculated (Napoli, Massetti, Brandani, Petralli, & Orlandini, 2016). Results can be further processed for energy simulation (Seyam, 2019). As such, simulation output tends to be in W/m^2 or kWh. Note that these are expressions of power over space or time and are not suitable for assessment for plant light duration. As covered in the previous section, the appropriate method for assessing light provision for plants should be to assess the amount of Photosynthetically Active Radiation (PAR) in the area and also to consider the duration of exposure. Therefore, some conversion is necessary.

After simulation, global solar radiation values will have to be converted to the Daily Light Integral (DLI) (Monteith, 1972). For tropical regions, the PAR can be estimated from global solar radiation based on the following equation (Tan & Ismail, 2016):

$$PAR = 1.867 * G, \tag{12}$$

where

PAR = Photosynthetically active radiation (mol/m^2)
G = Global solar radiation (MJ/m^2)

If simulated results of solar radiation are expressed as kilowatt-hour (kWh), it can be converted into Megajoules (MJ) based on the following conversion:

$$1 \ kWh = 3.6 \ MJ \tag{13}$$

If simulation results are cumulative, it will have to be divided by the number of simulated days to get the DLI:

$$DLI = (1.867 * G)/D, \tag{14}$$

• **Heat and energy**
Simulation of building energy consumption.
Changes in thermal transmission for
building envelope due to presence
of green walls or green roofs.

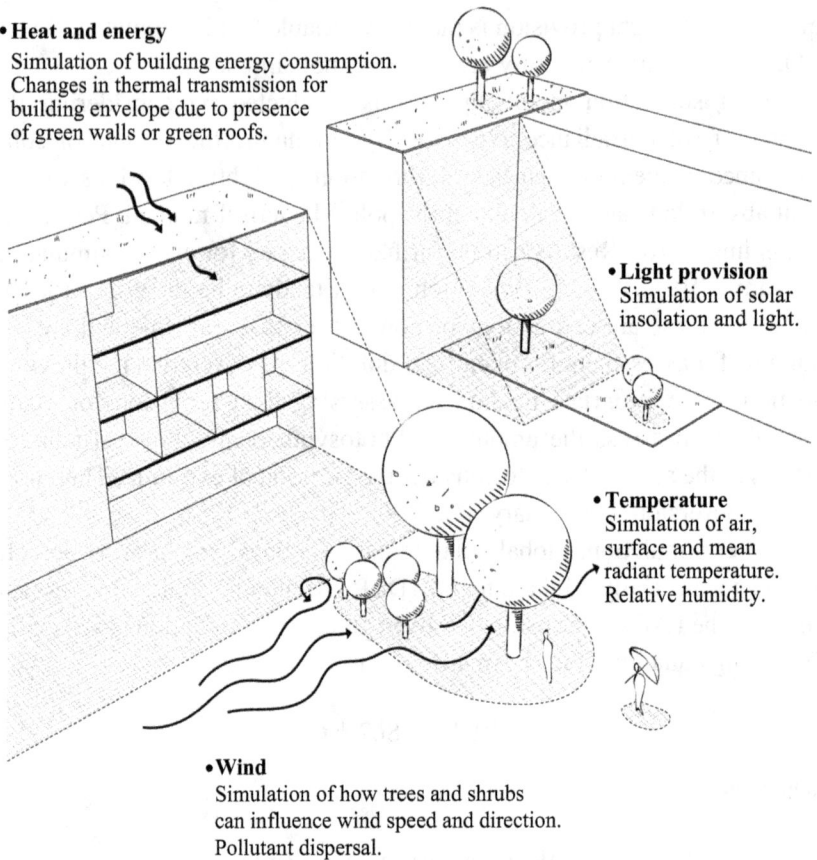

• **Light provision**
Simulation of solar
insolation and light.

• **Temperature**
Simulation of air,
surface and mean
radiant temperature.
Relative humidity.

• **Wind**
Simulation of how trees and shrubs
can influence wind speed and direction.
Pollutant dispersal.

Fig. 42.   Simulation of the impact of the environment on greenery and vice versa.

where

DLI = Daily light integral (mol/m^2/d)
G = Global solar radiation (MJ/m^2)
D = No. of simulated days

Simulating the impact of plants on the environment

Another common reason for conducting simulation is to estimate the impact of greenery on its surrounding environment (Fig. 42). This is relevant when designing to improve outdoor conditions or to minimise energy usage. It is an important step towards an iterative, evidence-based design framework for optimising ecosystem services provided by urban greenery.

Some examples include:

Temperature

Temperature simulations are useful for estimating potential reductions in temperature in the presence of greenery. Shade cast by tree canopies block solar radiation and significantly reduces heat exposure, lowering both the mean radiant temperature as well as surface temperature (Zhao, Chen, & Li, 2022). With sufficient canopy cover, even a single tree can provide significant reductions in surface temperature. Air temperature simulation also considers the cooling effect due to increased latent heat of vaporisation from evapotranspiration. Both physical (Teshnehdel, Akbari, Di Giuseppe, & Brown, 2020) and empirical models (Ignatius, Wong, & Jusuf, 2015) have been used for simulation.

Wind speed and direction

Trees, shrubs and hedges are large objects and may affect ventilation when placed in large quantities. Computational Fluid Dynamics (CFD) simulation can be used to estimate the impact of a planting scheme on the overall wind conditions (Kang, Kim, & Choi, 2020).

Energy consumption

Greenery can help to reduce energy consumption by reducing thermal transmission into buildings. Through simulation, the amount of energy saved can be quantified. Typically, the thermal properties of greenery (or greenery systems) will be appended as input parameters, changing the thermal properties of the building envelope (Wong, Tan, Tan, & Wong, 2009). In this case, greenery such as green walls or green roofs act as thermal insulation for the building. The reduction of air temperature due to the presence of greenery can also reduce cooling load for buildings

(Morakinyo, Lau, Ren, & Ng, 2018). Results from energy simulations can help promote the implementation of greenery to improve the thermal environment, leading to savings in energy and cost.

Common profiles

This section provides a reference for some commonly encountered profiles for light, temperature and water when embarking on field measurement. Measurements of lighting are mainly for ascertaining suitability for plant growth.

Solar radiation

In the tropical climate, the sun's path is quite consistent throughout the year. Peak temperature can be expected when the sun is at its zenith. Since there are no seasonal changes, peak temperature typically is around 12:00 hrs to 14:00 hrs. Peak solar irradiance can range from 700 W/m^2 to 1,000 W/m^2. Overcast sky conditions have a profile similar to that of clear sky conditions but smaller in magnitude (Fig. 43). Significant dips may indicate more clouds or a rain event.

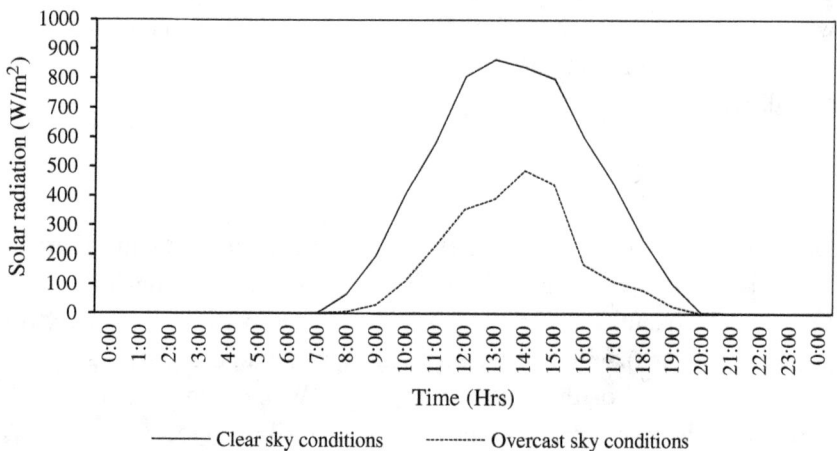

Fig. 43.   Typical diurnal outdoor profile for solar irradiance.

Photosynthetically Active Radiation

Photosynthetically Active Radiation (PAR) profiles in outdoor conditions will have the same profiles as solar irradiance, as the source for both variables are the same (the sun) (Fig. 44). The difference is that PAR is used specifically for estimating light provision for plants whereas solar irradiance is used for understanding general climatic conditions.

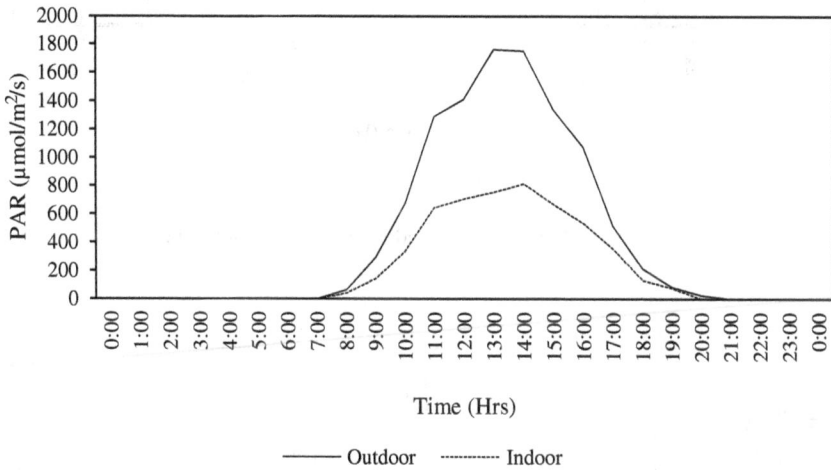

Time (Hrs)

———— Outdoor   ---------- Indoor

Fig. 44.   Typical diurnal outdoor profile for PAR.

For indoor environments where growth lights are installed, a regular box-shaped profile can be observed. PAR values will increase instantaneously when growth lights are on and will remain constant until they are turned off (Fig. 45). The amount of lighting provided may be akin to an outdoor shaded environment. In some cases, measurements may show a combination of regular box-shaped profiles as well as irregular peaks (Fig. 46). This signals the presence of both artificial lighting as well as natural daylighting. For instance, the area may be exposed to direct solar radiation for certain periods of the day from a glass façade. There is potential over-compensation of lighting for such scenarios, and the growth light provision period may be reduced accordingly.

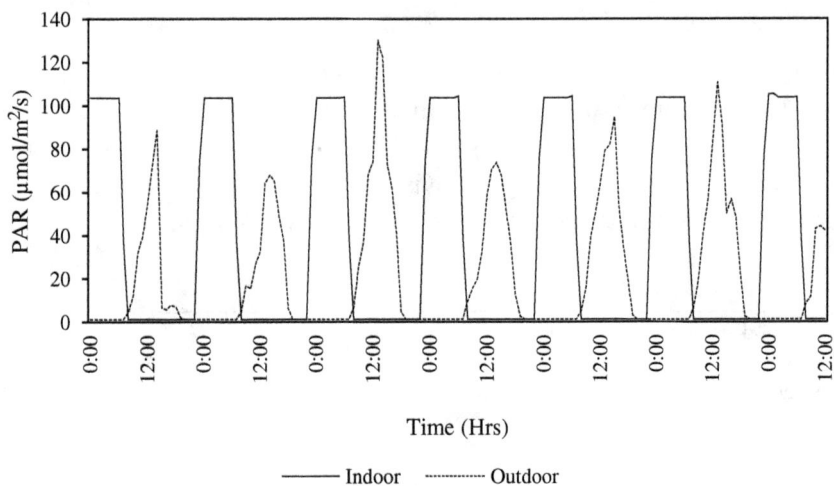

Fig. 45. Typical profile for PAR in indoor and outdoor shaded conditions.

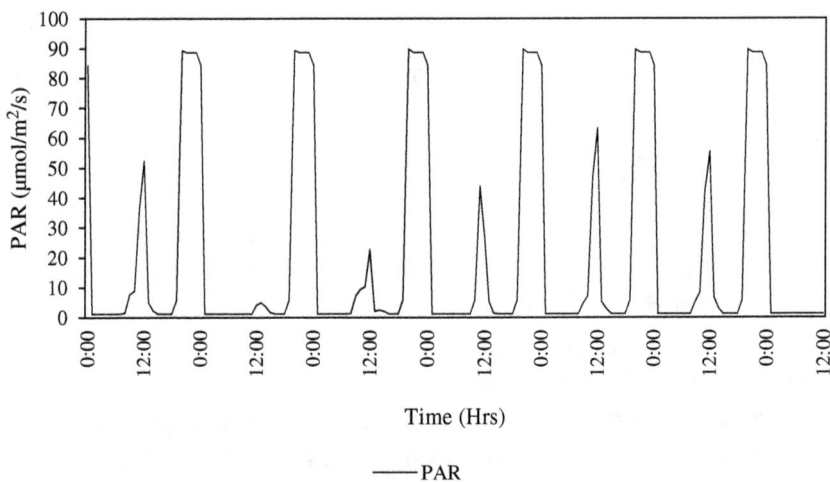

Fig. 46. Typical indoor profile for PAR with partial daylight exposure.

Temperature

Measurements of temperature are mainly used to estimate the impact of greenery on microclimatic conditions. For observations to be meaningful, the type of temperature measurement should first be defined.

Air temperature

Air temperature measurements can show how greenery can help cool the surrounding air. The cooling potential of greenery is chiefly due to the effects of shading and evapotranspiration (ET). Therefore, cooling is more pronounced during periods of active ET and high solar irradiance (Fig. 47). In the absence of sunlight, air temperature may be slightly higher with greenery as some heat is trapped in the canopy.

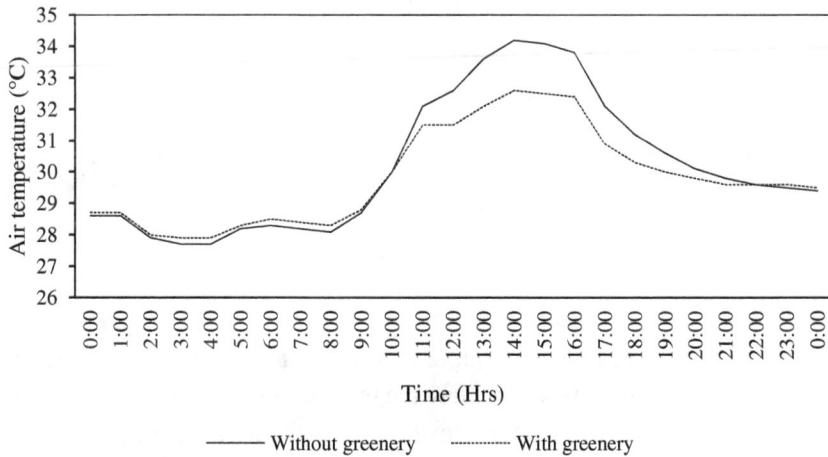

Fig. 47.   Typical diurnal outdoor profile for air temperature.

Mean radiant temperature

Measurements of mean radiant temperature (T_{mrt}) provide detailed information on the radiant qualities of the environment. Compared to air temperature, both magnitude and range are much larger. The diurnal profile of T_{mrt} will be similar to that of solar irradiance, since they are directly correlated. In the presence of trees, shade provided by canopies will reduce T_{mrt} drastically and significantly improve thermal comfort (Fig. 48).

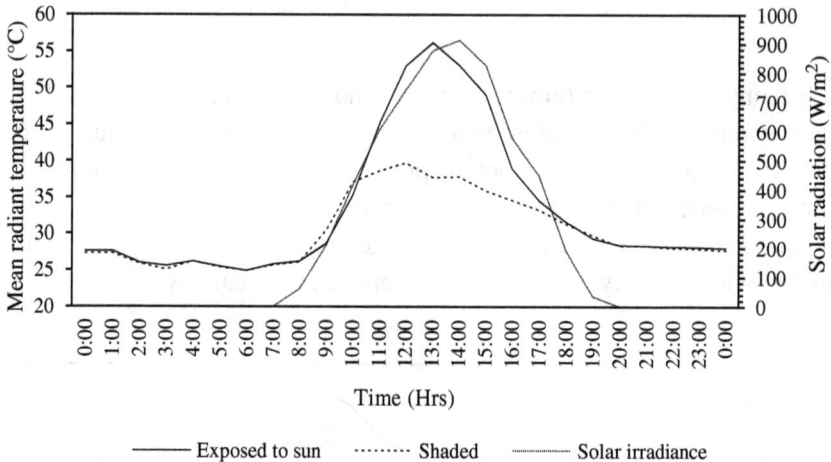

Fig. 48. Typical diurnal outdoor profile for mean radiant temperature.

Surface temperature

Surface temperature measurements can show how greenery can provide cooling by reducing the temperature of surfaces and thermal transmission into buildings. Figure 49 shows how the temperature of a concrete surface can be significantly reduced by covering it with an extensive green roof system. A reduction of more than 25°C in surface temperature can be observed during periods of high solar exposure.

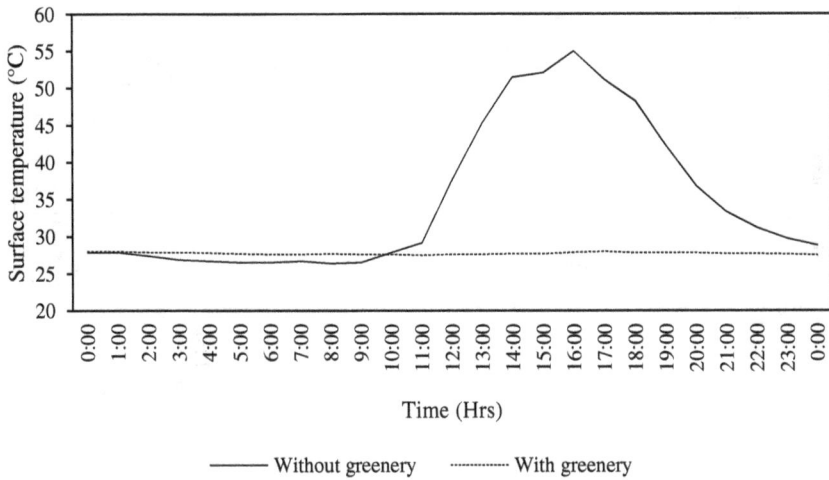

Fig. 49.   Typical diurnal outdoor profile for surface temperature.

Water content

Measurements of water content are used to estimate the amount of water in the soil or substrate, a critical component in maintaining good plant health. During irrigation, volumetric water content is expected to rise. When irrigation is interrupted, the volumetric water content level will slowly decrease over time. Care should be taken to ensure that values do not fall below 0.1 m^3/m^3, which should be considered the absolute minimum below which significant plant deterioration will ensue (Fig. 50). On the green wall, volumetric water content is dependent on the location of the sensors. For sensors placed closer to the irrigation drip emitters, higher fluctuations in water content can be observed. Readings tend to be more stable in the middle and lower portions of the green wall segments (Fig. 51).

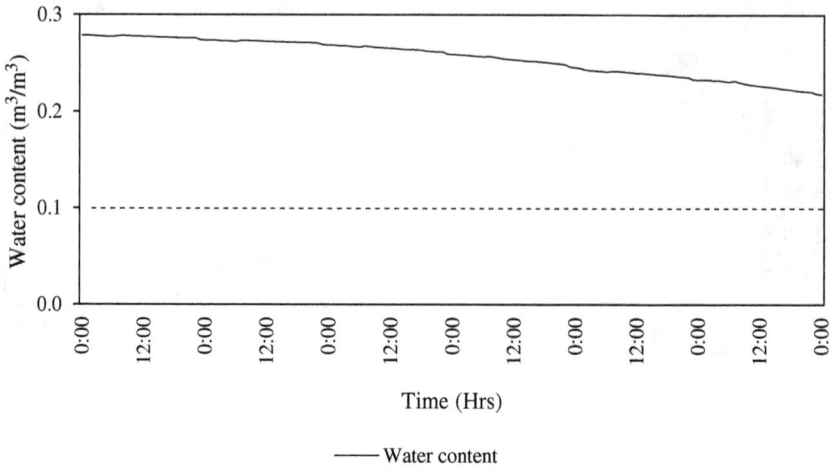

Fig. 50.   Typical indoor profile for volumetric water content in substrate on a green wall that has not received regular irrigation.

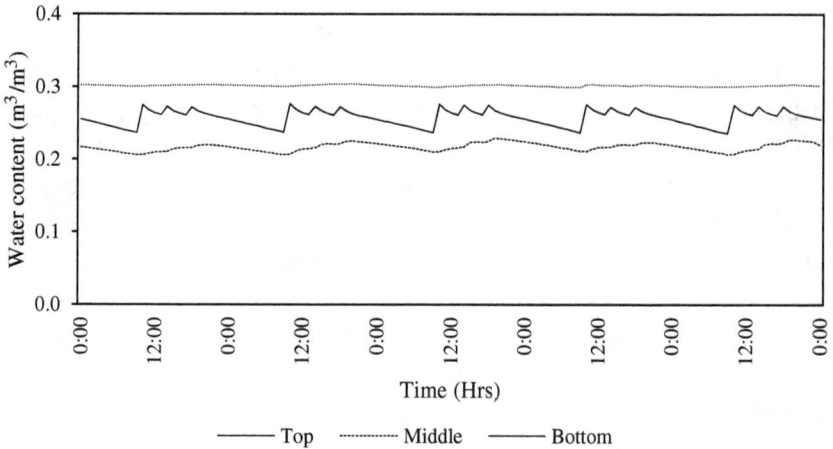

Fig. 51.   Typical indoor profile for volumetric water content in substrate on a green wall, showing differing profiles due to location of sensors.

Measurement flow chart

A flow chart for facilitating decision-making regarding measurement is shown in Fig. 52. It is recommended that users refer to the chart as well as information provided in this chapter to formulate and deploy a measurement plan.

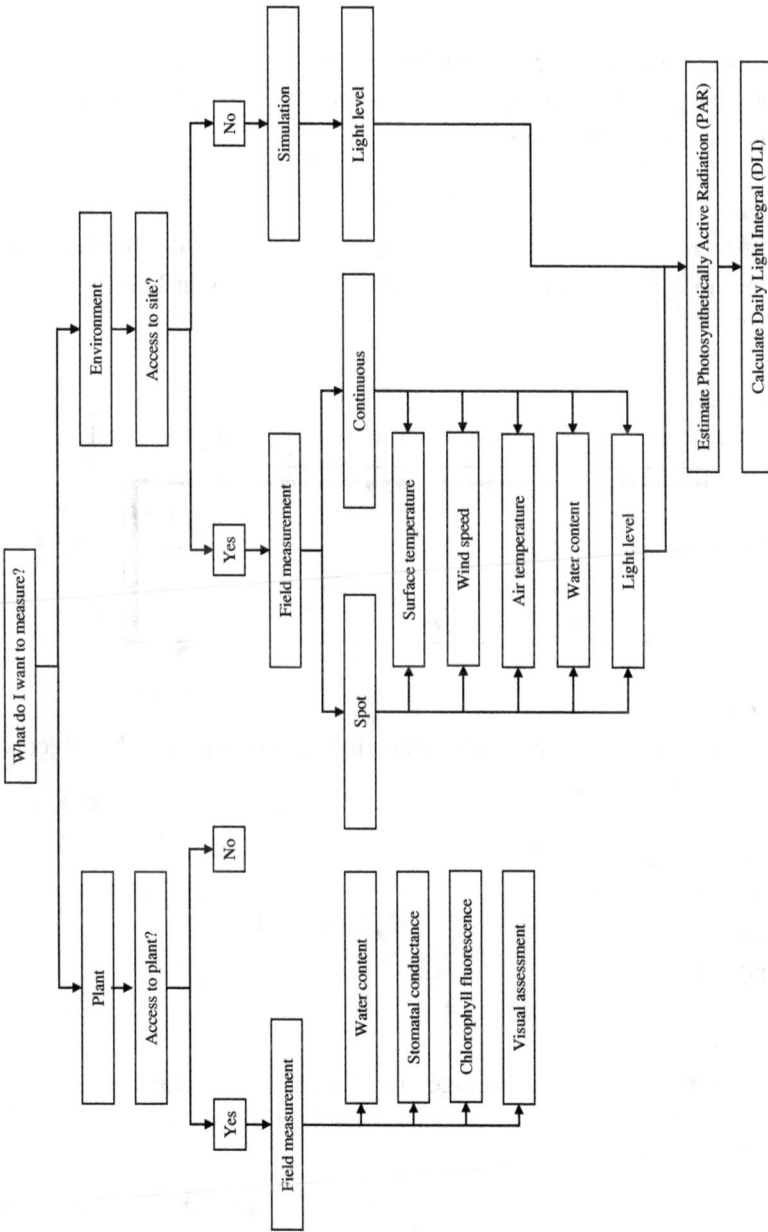

Fig. 52.   Flowchart for measurement.

Exercises

Question 1:
In-situ measurement of the lighting level over a period of 1 month showed an average light exposure of 1,250 Mlux per day. What group of plants is suitable for such lighting conditions?

Question 2:
You measure an outdoor area and get consistent PAR readings of 750 μmol/m^2/s for 4 hours a day. What is the Daily Light Integral (DLI) of the area?

Question 3:
Where should you place the soil moisture sensors on a green wall?

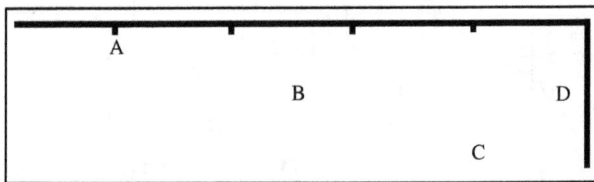

Question 4:
In a typical modelling exercise for simulating light provision, what objects should not be modelled?

A) Building blocks
B) Railings
C) Trees
D) Windows

Question 5:
Simulation results show an average cumulative yearly solar exposure of 800 kWh for the area of interest. What is the DLI of this area?

CHAPTER 6

Solutions to exercises

City-scale greenery

Question 1:

Biotope Area Factor (BAF)

$$= \frac{55 + 1,010 + 275 + 154}{60 + 220 + 1,110 + 210 + 550 + 250 + 800}$$

$$= \frac{1,494}{3,200}$$

$$= 0.467$$

Question 2:

Green Plot Ratio (GnPR)

Category	Subcategory	(A) LAI	(B) Canopy area	(C) Quantity	(A X B X C) Leaf area
Trees (no.)	Open canopy	2.5	60	6	900
	Intermediate canopy	3.0	60	12	2,160
	Dense canopy	4.0	60	43	10,320
Palms (no.)	Solitary	2.5	20	6	300
	Cluster	4.0	17	0	0
Shrubs (m²)	Monocot	3.5	—	64	224
	Dicot	4.5	—	20	90
Turf (m²)	—	2.0	—	130	260
Green walls (m²)	—	2.0	—	55	110

Green Plot Ratio (GnPR)

$$= \frac{900 + 2,160 + 10,320 + 300 + 224 + 90 + 260 + 110}{4,500}$$

$$= \frac{14,364}{4,500}$$

$$= 3.192$$

Plant-scale greenery

Question 1:

Measurement period = 1 month
Average daily light level = 1,250 Mlux

Step 1: Convert Lux to PAR

$$PAR = 0.021 * L \tag{9}$$

where

PAR = Photosynthetically Active Radiation (mol/m^2)
L = Lux (Mlux)
PAR = 0.021 * 1,250 = 26.25 mol/m^2

Since 1,250 Mlux is already the <u>daily</u> light level, it is showing the amount of PAR exposure per day, which is equivalent to the DLI.

Therefore, the DLI of area = 26.25 mol/m^2/d

Referring to Table 10, this area is suitable for full sun plants.

Lighting requirements	DLI range (mol/m^2/d)
Full sun	> 12.0
Partial shade	> 6.0–12.0
Full shade	3.0–6.0

Question 2:

Measurement period = 4 hrs a day
PAR = 750 μmol/m²/s

Step 1: Find the DLI

$$DLI = PAR * 0.0036 * T, \tag{10}$$

where

DLI = Daily Light Integral (mol/m²/d)
PAR = Photosynthetically Active Radiation (μmol/m²/s)
T = Duration of exposure (Hr)
DLI of area = $750 \times 0.0036 \times 4 = 10.8$ mol/m²/d

Question 3:

Answer = B

Distribution of water is not homogenous across the green wall.

A is too close to an emitter.
C is too low, a possible location, but not as good as B.
D is too close to the edge of the wall.

Question 4:

Answer = B

Modelled objects should be relevant and have a significant influence on simulated parameters.

Building blocks create shading that can drastically alter the light provision for plants as well as thermal conditions.

Tree canopies provide shade and reduce temperature via evaporative cooling. Trees also influence the wind flow.

Windows affect lighting into buildings, leading to changes in light provision as well as energy consumption.

Railings do not affect environmental conditions significantly.

Question 5:

Simulation duration = 1 year = 365 days
Average cumulative yearly solar exposure = 800 kWh

Step 1: Convert kWh to MJ

$$1 \ kWh = 3.6 \ MJ \tag{13}$$

800 kWh = 3.6 x 800 = 2,880 MJ

Step 2: Calculate the DLI

$$DLI = (1.867 * G)/D, \tag{14}$$

where

DLI = Daily light integral (mol/m^2/d)
G = Global solar radiation (kWh)
D = No. of simulated days
DLI = 1.867 x 2880 / 365 = 14.7 mol/m^2/d

Instructions for Supplementary Material

The author has prepared additional resources for readers of *Urban Greening Techniques* on www.ugl.sg/book. To view these supplementary material, please enter the password "ugl".

Appendix

List of landscape metrics (McGarigal, 1995)

Category	Metric	Scale
Core area	Core area	P
	Number of core areas	P
	Core area index	P
	Core area percentage of landscape	C
	Total core area	C/L
	Number of core areas	C/L
	Core area density	C/L
	Mean core area per patch	C/L
	Patch core area standard deviation	C/L
	Patch core area coefficient of variation	C/L
	Mean area per disjunct core	C/L
	Disjunct core area standard deviation	C/L
	Disjunct core area coefficient of variation	C/L
	Total core area index	C/L
	Mean core area index	C/L
Area	Area	P
	Landscape similarity index	P
	Class area	C
	Percentage of landscape	C
	Total landscape area	C/L
	Largest patch index	C/L

(*Continued*)

(*Continued*)

Category	Metric	Scale
Shape	Shape index	P
	Fractal dimension	P
	Landscape shape index	C/L
	Mean shape index	C/L
	Area-weighted mean shape index	C/L
	Double log fractal dimension	C/L
	Mean patch fractal dimension	C/L
	Area-weighted mean patch fractal dimension	C/L
Edge	Perimeter	P
	Edge contrast index	P
	Total edge	C/L
	Edge density	C/L
	Contrast-weighted edge density	C/L
	Total edge contrast index	C/L
	Mean edge contrast index	C/L
	Area-weighted mean edge contrast index	C/L
Patch density, size and variability	Number of patches	C/L
	Patch density	C/L
	Mean patch size	C/L
	Patch size standard deviation	C/L
	Patch size coefficient of variation	C/L
Nearest neighbour	Nearest neighbour distance	P
	Proximity index	P
	Mean nearest neighbour distance	C/L
	Nearest neighbour standard deviation	C/L
	Nearest neighbour coefficient of variation	C/L
	Mean proximity index	C/L
Diversity	Shannon's diversity index	L
	Simpson's diversity index	L
	Modified Simpson's diversity index	L
	Patch richness	L
	Patch richness density	L
	Relative patch richness	L

(*Continued*)

Category	Metric	Scale
	Shannon's evenness index	L
	Simpson's evenness index	L
	Modified Simpson's evenness index	L
Contagion and interspersion	Interspersion and Juxtaposition index	C/L
	Contagion index	L

P = Patch

C = Class

L = Landscape

Bibliography

Ade, R., & Rehm, M. (2020). The unwritten history of green building rating tools: A personal view from some of the "founding fathers". *Building Research & Information, 48*(1), 1–17.

Akbar, T. A., Hassan, Q. K., Ishaq, S., Batool, M., Butt, H. J., & Jabbar, H. (2019). Investigative spatial distribution and modelling of existing and future urban land changes and its impact on urbanization and economy. *Remote Sensing, 11*(2), 105.

Alexandri, E., & Jones, P. (2008). Temperature decreases in an urban canyon due to green walls and green roofs in diverse climates. *Building and Environment, 43*(4), 480–493.

Andersson, E., Barthel, S., & Ahrné, K. (2007). Measuring social-ecological dynamics behind the generation of ecosystem services. *Ecological Applications, 17*(5), 1267–1278.

Auckland Council. (2018). *Auckland Plan 2050.* Auckland Council Te Kaunihera o Tāmaki Makaurau.

Auckland Council. (2019). *Auckland's Urban Ngahere (Forest) Strategy.* https://www.aucklandcouncil.govt.nz/plans-projects-policies-reports-bylaws/our-plans-strategies/topic-based-plans-strategies/environmental-plans-strategies/Documents/urban-ngahere-forest-strategy.pdf

Azkorra, Z., Pérez, G., Coma, J., Cabeza, L. F., Bures, S., Álvaro, J. E., … Urrestarazu, M. (2015). Evaluation of green walls as a passive acoustic insulation system for buildings. *Applied Acoustics, 89*, 46–56. https://doi.org/10.1016/j.apacoust.2014.09.010

Baligar, V., Fageria, N., & He, Z. (2001). Nutrient use efficiency in plants. *Communications in Soil Science and Plant Analysis, 32*(7–8), 921–950.

Banks, J. (1998). Principles of simulation. In *Handbook of Simulation: Principles, Methodology, Advances, Applications, and Practice*, 3–30.

Bass, B., Liu, K., & Baskaran, B. (2003). Evaluating rooftop and vertical gardens as an adaptation strategy for urban areas. Technical Report. National Research Council Canada. https://doi.org/10.4224/20386110

Bassuk, N. (2008). CU-Structural soil: An update after more than a decade of use in the urban environment. *The Society of Municipal Arborists (City-Trees)*.

BCA. (2022a). GM: 2021 Health & wellbeing. https://www1.bca.gov.sg/docs/default-source/bca-awards-2020/20220419_healthwellbeing_simplified_r1-2.pdf?sfvrsn=d8f5e495_0

BCA. (2022b). GM: 2021 Resilience. https://www1.bca.gov.sg/docs/default-source/docs-corp-buildsg/sustainability/20210907_resilience_simplified.pdf

BCA. (2022c). GM: 2021 Certification standard. https://www1.bca.gov.sg/docs/default-source/docs-corp-buildsg/sustainability/20211028_certification-standard_r1-1.pdf

Beatley, T. (2016). Singapore city, Singapore: City in a garden. *Handbook of Biophilic City Planning and Design*, 51–66.

Belcher, R. N., & Chisholm, R. A. (2018). Tropical vegetation and residential property value: A hedonic pricing analysis in Singapore. *Ecological Economics, 149*, 149–159.

Bengtsson, L. (2005). Peak flows from thin sedum-moss roof. *Hydrology Research, 36*(3), 269–280.

Berardi, U., GhaffarianHoseini, A., & GhaffarianHoseini, A. (2014). State-of-the-art analysis of the environmental benefits of green roofs. *Applied Energy, 115*, 411–428.

Berghöfer, A., Mader, A., Patrickson, S., Calcaterra, E., Smit, J., Blignaut, J., ... van Zyl, H. (2011). TEEB Manual for cities: Ecosystem services in urban management. *The Economics of Ecosystems and Biodiversity*. https://teebweb.org/publications/other/teeb-cities/

Berndtsson, J. C. (2010). Green roof performance towards management of runoff water quantity and quality: A review. *Ecological Engineering, 36*(4), 351–360.

Berndtsson, J. C., Emilsson, T., & Bengtsson, L. (2006). The influence of extensive vegetated roofs on runoff water quality. *Science of the Total Environment, 355*(1–3), 48–63.

Bianchini, F., & Hewage, K. (2012a). How "green" are the green roofs? Lifecycle analysis of green roof materials. *Building and Environment, 48*, 57–65.

Bianchini, F., & Hewage, K. (2012b). Probabilistic social cost-benefit analysis for green roofs: A lifecycle approach. *Building and Environment, 58*, 152–162.

Biljecki, F., Ledoux, H., & Stoter, J. (2016). An improved LOD specification for 3D building models. *Computers, Environment and Urban Systems, 59*, 25–37.

Bombieri, G., Delgado, M. d. M., Russo, L. F., Garrote, P. J., López-Bao, J. V., Fedriani, J. M., & Penteriani, V. (2018). Patterns of wild carnivore attacks on humans in urban areas. *Scientific Reports, 8*(1), 1–9.

Bonan, G. (2015). *Ecological Climatology: Concepts and Applications* (3rd ed.). Cambridge University Press, Cambridge.

Boo, C. M., Omar-Hor, K., Ou-Yang, C. L., & Ng, C. K. (2003). 1001 garden plants in Singapore. National Parks Board.

Bratman, G. N., Anderson, C. B., Berman, M. G., Cochran, B., De Vries, S., Flanders, J., … Hartig, T. (2019). Nature and mental health: An ecosystem service perspective. *Science Advances, 5*(7), eaax0903.

BRE. (2020). BREEAM in-use international technical manual. BRE Global Limited, United Kingdom.

Brears, R. C. (2018). *Blue and Green Cities: The Role of Blue-Green Infrastructure in Managing Urban Water Resources*. Springer.

Buccolieri, R., Santiago, J.-L., Rivas, E., & Sanchez, B. (2018). Review on urban tree modelling in CFD simulations: Aerodynamic, deposition and thermal effects. *Urban Forestry & Urban Greening, 31*, 212–220.

Buchmann, S. L., & Nabhan, G. P. (2012). *The Forgotten Pollinators*. Island Press.

Bush, J., Ashley, G., Foster, B., & Hall, G. (2021). Integrating green infrastructure into urban planning: Developing Melbourne's green factor tool. *Urban Planning and Green Infrastructure, 6*(1), 20–31.

Bustamante, W., Uribe, D., Vera, S., & Molina, G. (2017). An integrated thermal and lighting simulation tool to support the design process of complex fenestration systems for office buildings. *Applied Energy, 198*, 36–48.

Calloway, E. (2019). The long road to the "all" of HIAP. *Chronicles of Health Impact Assessment, 4*(1), 45–66.

Cameron, R. W., Blanuša, T., Taylor, J. E., Salisbury, A., Halstead, A. J., Henricot, B., & Thompson, K. (2012). The domestic garden — Its contribution to urban green infrastructure. *Urban Forestry & Urban Greening, 11*(2), 129–137.

Carter, T., & Fowler, L. (2008). Establishing green roof infrastructure through environmental policy instruments. *Environmental Management, 42*(1), 151–164.

Carter, T., & Jackson, C. R. (2007). Vegetated roofs for stormwater management at multiple spatial scales. *Landscape and Urban Planning, 80*(1–2), 84–94.

Champagne, C. L., & Aktas, C. B. (2016). Assessing the resilience of LEED certified green buildings. *Procedia Engineering, 145*, 380–387.

Chan, F. K. S., Griffiths, J. A., Higgitt, D., Xu, S., Zhu, F., Tang, Y.-T., ... Thorne, C. R. (2018). "Sponge City" in China — a breakthrough of planning and flood risk management in the urban context. *Land Use Policy, 76*, 772–778.

Chan, L., Hillel, O., Elmqvist, T., Werner, P., Holman, N., Mader, A., & Calcaterra, E. (2014). User's manual on the Singapore index on cities' biodiversity (also known as the City Biodiversity Index). National Parks Board, Singapore.

Chanchitpricha, C., & Fischer, T. B. (2022). The role of impact assessment in the development of urban green infrastructure: A review of EIA and SEA practices in Thailand. *Impact Assessment and Project Appraisal, 40*(3), 191–201.

Chang, C.-R., Li, M.-H., & Chang, S.-D. (2007). A preliminary study on the local cool-island intensity of Taipei city parks. *Landscape and Urban Planning, 80*(4), 386–395.

Chang, X., Alderson, P. G., & Wright, C. J. (2008). Solar irradiance level alters the growth of basil (*Ocimum basilicum L.*) and its content of volatile oils. *Environmental and Experimental Botany, 63*(1–3), 216–223.

Chiang, K., & Tan, A. (2009). Vertical greenery for the tropics. Centre for Urban Greenery and Ecology, Singapore.

Choi, J., & Kim, G. (2022). History of Seoul's parks and green space policies: Focusing on policy changes in urban development. *Land, 11*(4), 474.

Churkina, G., Grote, R., Butler, T. M., & Lawrence, M. (2015). Natural selection? Picking the right trees for urban greening. *Environmental Science & Policy, 47*, 12–17.

City of Helsinki Environment Centre. (2016). Developing the city of Helsinki green factor method. https://www.integratedstormwater.eu/sites/www.integratedstormwater.eu/files/report_summary_developing_a_green_factor_tool_for_the_city_of_helsinki.pdf

CityLAB, B. (2021). Gieß den Kiez. https://www.giessdenkiez.de/

Civil Engineering and Development Department. (2019). Greening Master Plan for urban areas. Hong Kong Special Administrative Region.

Clark, C., Adriaens, P., & Talbot, F. B. (2008). Green roof valuation: a probabilistic economic analysis of environmental benefits. *Environmental Science & Technology, 42*(6), 2155–2161.

Cohen, A. J., Anderson, H. R., Ostro, B., Pandey, K. D., Krzyzanowski, M., Künzli, N., ... Samet, J. M. (2004). Urban air pollution. In *Comparative*

Quantification of Health Risks: Global and Regional Burden of Disease Attributable to Selected Major Risk Factors, 2, 1353–1433.

Coma, J., Perez, G., de Gracia, A., Burés, S., Urrestarazu, M., & Cabeza, L. F. (2017). Vertical greenery systems for energy savings in buildings: A comparative study between green walls and green facades. *Building and Environment, 111*, 228–237.

CUGE. (2015). Sustainable landscape. https://www.nparks.gov.sg/-/media/cuge/ ebook/sustainable-landscape/sustainable-landscape.pdf

D'amato, G. (2000). Urban air pollution and plant-derived respiratory allergy. *Clinical & Experimental Allergy 30*, 628–636.

Dahlan, M., Faisal, B., Chaeriyah, S., Hutriani, I., & Amelia, M. (2022). The challenges of implementing green factors in urban greening schemes in Indonesia. Paper presented at the IOP Conference Series: Earth and Environmental Science.

Darlington, A., Chan, M., Malloch, D., Pilger, C., & Dixon, M. A. (2000). The biofiltration of indoor air: Implications for air quality. *Indoor Air, 10*(1), 39–46. https://doi.org/10.1034/j.1600-0668.2000.010001039.x

Datta, R. (2016). Community garden: A bridging program between formal and informal learning. *Cogent Education, 3*(1), 1177154.

David, V., Michelle, P., Neil, P., & Kevin, L. (2016). Urban landscape maintenance in Singapore: Special report on landscape productivity management. National Parks Board, Singapore.

Dawson, R., Wyckmans, A., Heidrich, O., Köhler, J., Dobson, S., & Feliu, E. (2014). Understanding cities: Advances in integrated assessment of urban sustainability. Centre for Earth Systems Engineering Research (CESER), Newcastle University.

de Jong, K., Albin, M., Skärbäck, E., Grahn, P., & Björk, J. (2012). Perceived green qualities were associated with neighborhood satisfaction, physical activity, and general health: Results from a cross-sectional study in suburban and rural Scania, southern Sweden. *Health Place, 18*(6), 1374–1380. doi:10.1016/j.healthplace.2012.07.001

de Vries, G. (2021). Green infrastructures in Amsterdam: A case study on the governance challenges of green infrastructure implementation. MA Thesis, University of Utrecht.

del Carmen Redondo-Bermúdez, M., Gulenc, I. T., Cameron, R. W., & Inkson, B. J. (2021). "Green barriers" for air pollutant capture: leaf micromorphology as a mechanism to explain plants capacity to capture particulate matter. *Environmental Pollution, 288*, 117809.

Delmas, M. A., & Burbano, V. C. (2011). The drivers of greenwashing. *California Management Review, 54*(1), 64–87.

Deng, L., Luo, H., Ma, J., Huang, Z., Sun, L.-X., Jiang, M.-Y., ... Greening, U. (2020). Effects of integration between visual stimuli and auditory stimuli on restorative potential and aesthetic preference in urban green spaces. US Department of Agriculture, p. 126702.

Department of Planning and Sustainability, Amsterdam (2021). Policy: Green space. https://www.amsterdam.nl/en/policy/policy-green-space/

Detroit Greenways Coalition. (2007). Maps and data. https://detroitgreenways. org/maps-and-data/

Dorais, M. (2003). The use of supplemental lighting for vegetable crop production: light intensity, crop response, nutrition, crop management, cultural practices. Paper presented at the Canadian Greenhouse Conference.

Ebenezer, H. (1898). *Tomorrow: A Peaceful Path to Real Reform*. Swan Sonnenschein & Co., London, pp. 6–7.

Ellis, J. A., Walter, A., Tooker, J. F., Ginzel, M. D., Reagel, P. F., Lacey, E. S., ... Hanks, L. M. (2005). Conservation biological control in urban landscapes: Manipulating parasitoids of bagworm (*Lepidoptera: Psychidae*) with flowering forbs. *Biological Control, 34*(1), 99–107.

Ernstson, H., Van der Leeuw, S. E., Redman, C. L., Meffert, D. J., Davis, G., Alfsen, C., & Elmqvist, T. (2010). Urban transitions: On urban resilience and human-dominated ecosystems. *Ambio, 39*(8), 531–545.

Escobedo, F. J., Kroeger, T., & Wagner, J. E. (2011). Urban forests and pollution mitigation: Analyzing ecosystem services and disservices. *Environmental Pollution, 159*(8–9), 2078–2087.

Esmail, B. A., Cortinovis, C., Suleiman, L., Albert, C., Geneletti, D., Mörtberg, U. J. U. F., & Greening, U. (2022). Greening cities through urban planning: A literature review on the uptake of concepts and methods in Stockholm. *Urban Forestry & Urban Greening*, 127584.

Fahrig, L., Baudry, J., Brotons, L., Burel, F. G., Crist, T. O., Fuller, R. J., ... Martin, J. L. (2011). Functional landscape heterogeneity and animal biodiversity in agricultural landscapes. *Ecology Letters, 14*(2), 101–112.

Fang, C.-F., & Ling, D.-L. (2003). Investigation of the noise reduction provided by tree belts. *Landscape and Urban Planning, 63*(4), 187–195.

Feng, H., & Hewage, K. N. (2018). Economic benefits and costs of green roofs. In *Nature Based Strategies for Urban and Building Sustainability*. Elsevier, pp. 307–318.

Feng, Y., & Tan, P. Y. (2017). Imperatives for greening cities: A historical perspective. *Greening Cities: Forms and Functions*, pp. 41–70.

Fiorello, I., Del Dottore, E., Tramacere, F., & Mazzolai, B. (2020). Taking inspiration from climbing plants: methodologies and benchmarks — a review. *Bioinspiration & Biomimetics, 15*(3), 031001.

Fischer, J., Stott, J., & Law, B. S. (2010). The disproportionate value of scattered trees. *Biological Conservation, 143*(6), 1564–1567.

Forestry and Recreation Division, Toronto (2013). Sustaining and expanding the urban forest: Toronto's strategic forest management plan. Unpublished report, City of Toronto, Parks, Forestry and Recreation Division.

Forman, R. T. (1995). Some general principles of landscape and regional ecology. *Landscape Ecology, 10*(3), 133–142.

Forrest, M., & Konijnendijk, C. (2005). A history of urban forests and trees in Europe. In *Urban Forests and Trees: A Reference Book*, pp. 23–48.

Foster, J., Lowe, A., & Winkelman, S. (2011). The value of green infrastructure for urban climate adaptation. *Center for Clean Air Policy, 750*(1), 1–52.

Fu, G., Zhang, C., Hall, J. W., & Butler, D. (2023). Are sponge cities the solution to China's growing urban flooding problems? *Wiley Interdisciplinary Reviews: Water, 10*(1), e1613.

Gómez-Baggethun, E., Gren, Å., Barton, D. N., Langemeyer, J., McPhearson, T., O'Farrell, P., ... Kremer, P. (2013). Urban ecosystem services. In *Urbanization, Biodiversity and Ecosystem Services: Challenges and Opportunities*. Springer, Dordrecht, pp. 175–251.

Gopalan, R., & Radhakrishna, S. (2022). Cities and biodiversity: Hidden connections between the built form and life. In *Blue-Green Infrastructure Across Asian Countries*. Springer, pp. 141–162.

Grabosky, J., Bassuk, N., & Trowbridge, P. (2002). Structural soils: A new medium to allow urban trees to grow in pavement. LATIS Cornell University, New York.

Grant, G. (2017). Greater London Authority / Urban greening factor for London. The Ecology Consultancy.

Grantz, D., Garner, J., & Johnson, D. (2003). Ecological effects of particulate matter. *Environment International, 29*(2–3), 213–239.

Gratani, L., & Varone, L. (2007). Plant crown traits and carbon sequestration capability by *Platanus hybrida Brot.* in Rome. *Landscape and Urban Planning, 81*(4), 282–286.

Gratani, L., Varone, L., & Bonito, A. (2016). Carbon sequestration of four urban parks in Rome. *Urban Forestry & Urban Greening, 19*, 184–193.

Groff, F. R., & McCord, E. (2012). The role of neighborhood parks as crime generators. *Security Journal, 25*(1), 1–24.

Gromke, C., Blocken, B., Janssen, W., Merema, B., van Hooff, T., & Timmermans, H. (2015). CFD analysis of transpirational cooling by vegetation: Case study for specific meteorological conditions during a heat wave in Arnhem, Netherlands. *Building and Environment, 83*, 11–26.

Grzędzicka, E. (2019). Is the existing urban greenery enough to cope with current concentrations of $PM_{2.5}$, PM_{10} and CO_2? *Atmospheric Pollution Research, 10*(1), 219–233.

Guo, X., Gao, Z., Buccolieri, R., Zhang, M., & Shen, J. (2021). Effect of greening on pollutant dispersion and ventilation at urban street intersections. *Building and Environment, 203*, 108075.

Halverson, N. (2004). Review of constructed subsurface flow vs. surface flow wetlands. University of North Texas. https://digital.library.unt.edu/ark:/67531/metadc782197/m2/1/high_res_d/835229.pdf

Han, Y., Lee, J., Haiping, G., Kim, K.-H., Wanxi, P., Bhardwaj, N., ... Brown, R. J. (2022). Plant-based remediation of air pollution: A review. *Journal of Environmental Management, 301*, 113860.

Harish, V., & Kumar, A. (2016). A review on modeling and simulation of building energy systems. *Renewable and Sustainable Energy Reviews, 56*, 1272–1292.

Hartig, J. H., Scott, T., Gell, G., & Berk, K. (2022). Reconnecting people to the Detroit River — A transboundary effort. *Aquatic Ecosystem Health & Management, 25*(1), 27–38.

He, Y., Lin, E. S., Tan, C. L., Yu, Z., Tan, P. Y., & Wong, N. H. (2021). Model development of Roof Thermal Transfer Value (RTTV) for green roof in tropical area: A case study in Singapore. *Building and Environment, 203*, 108101.

Herrera, M., Natarajan, S., Coley, D. A., Kershaw, T., Ramallo-González, A. P., Eames, M., ... Wood, M. (2017). A review of current and future weather data for building simulation. *Building Services Engineering Research and Technology, 38*(5), 602–627.

Hesselbarth, M. H., Sciaini, M., With, K. A., Wiegand, K., & Nowosad, J. (2019). Landscapemetrics: An open source R tool to calculate landscape metrics. *Ecography, 42*(10), 1648–1657.

Ho, D. K. H. (2020). *Greening the Urban Habitat: A Quantitative and Empirical Approach*: World Scientific Publishing, Singapore.

Homes, S. (2022). LEED hits 100,000 certified projects. https://www.usgbc.org/articles/leed-hits-100000-certified-projects

Hsu, C.-B., Hsieh, H.-L., Yang, L., Wu, S.-H., Chang, J.-S., Hsiao, S.-C., ... Lin, H.-J. (2011). Biodiversity of constructed wetlands for wastewater treatment. *Ecological Engineering, 37*(10), 1533–1545.

Huang, Z., Tan, C. L., Lu, Y., & Wong, N. H. (2021). Holistic analysis and prediction of life cycle cost for vertical greenery systems in Singapore. *Building and Environment, 196*, 107735.

Hwang, Y. H., & Roscoe, C. J. (2017). Preference for site conservation in relation to on-site biodiversity and perceived site attributes: An on-site survey of unmanaged urban greenery in a tropical city. *Urban Forestry & Urban Greening, 28*, 12–20.

Hwang, Y. H., Yue, Z. E. J., & Tan, Y. C. (2017). Observation of floristic succession and biodiversity on rewilded lawns in a tropical city. *Landscape Research, 42*(1), 106–119.

Ichihara, K., & Cohen, J. P. (2011). New York City property values: What is the impact of green roofs on rental pricing? *Letters in Spatial and Resource Sciences, 4*(1), 21–30.

Ignatius, M., Wong, N. H., & Jusuf, S. K. (2015). Urban microclimate analysis with consideration of local ambient temperature, external heat gain, urban ventilation, and outdoor thermal comfort in the tropics. *Sustainable Cities and Society, 19*, 121–135.

Jiang, Y., Menz, S., & Peric, A. (2023). Urban greenery as a tool to enhance social integration? A case study of Altstetten-Albisrieden, Zürich. In *Sustainable Built Environment*. IntechOpen.

Jim, C. Y. (1998). Urban soil characteristics and limitations for landscape planting in Hong Kong. *Landscape and Urban Planning, 40*(4), 235–249.

Jim, C. Y. (2004). Green-space preservation and allocation for sustainable greening of compact cities. *Cities, 21*(4), 311–320.

Jimenez, R. B., Lane, K. J., Hutyra, L. R., & Fabian, M. P. (2022). Spatial resolution of Normalized Difference Vegetation Index and greenness exposure misclassification in an urban cohort. *Journal of Exposure Science & Environmental Epidemiology, 32*(2), 213–222.

Jorgensen, A., & Anthopoulou, A. (2007). Enjoyment and fear in urban woodlands — Does age make a difference? *Urban Forestry & Urban Greening, 6*(4), 267–278.

Juhola, S. (2018). Planning for a green city: The Green Factor tool. *Urban Forestry & Urban Greening, 34*, 254–258.

Kang, G., Kim, J.-J., & Choi, W. (2020). Computational fluid dynamics simulation of tree effects on pedestrian wind comfort in an urban area. *Sustainable Cities and Society, 56*, 102086.

Kazemi, F., & Mohorko, R. (2017). Review on the roles and effects of growing media on plant performance in green roofs in world climates. *Urban Forestry & Urban Greening, 23*, 13–26.

Keeley, M. (2011). The Green Area Ratio: An urban site sustainability metric. *Journal of Environmental Planning and Management, 54*(7), 937–958.

Keeley, M., Benton-Short, L., Keeley, M., & Benton-Short, L. (2019). Urban green space. In *Urban Sustainability in the US: Cities Take Action*, pp. 239–279.

Keten, A., Eroglu, E., Kaya, S., & Anderson, J. T. (2020). Bird diversity along a riparian corridor in a moderate urban landscape. *Ecological Indicators, 118*, 106751.

Khan, M. M., Akram, M. T., Janke, R., Qadri, R. W. K., Al-Sadi, A. M., & Farooque, A. A. (2020). Urban horticulture for food secure cities through and beyond COVID-19. *Sustainability, 12*(22), 9592.

Khuong, P. M., McKenna, R., & Fichtner, W. (2019). Analyzing drivers of renewable energy development in Southeast Asia countries with correlation and decomposition methods. *Journal of Cleaner Production, 213*, 710–722.

Kim, K., Kim, S., & Park, C.-Y. (2020). Food Security in Asia and the Pacific amid the COVID-19 Pandemic. Asian Development Bank.

Kirkham, M. B. (2014). *Principles of Soil and Plant Water Relations*. Academic Press.

KL City Hall. (2021). Kuala Lumpur Climate Action Plan 2050. City Planning Department. https://www.dbkl.gov.my/wp-content/uploads/2021/07/C40_KLCAP2050_viewing-only-MR-single_compressed.pdf

Kong, F., Chen, J., Middel, A., Yin, H., Li, M., Sun, T., … Zhou, K. (2022). Impact of 3-D urban landscape patterns on the outdoor thermal environment: A modelling study with SOLWEIG. *Computers, Environment and Urban Systems, 94*, 101773.

Konijnendijk, C. C., Annerstedt, M., Nielsen, A. B., & Maruthaveeran, S. J. (2013). Benefits of urban parks. International Federation of Parks and Recreation Administration. https://worldurbanparks.org/images/Newsletters/IfpraBenefitsOfUrbanParks.pdf

Korczynski, P. C., Logan, J., & Faust, J. E. (2002). Mapping monthly distribution of daily light integrals across the contiguous United States. *HortTechnology, 12*(1), 12–16.

Krivtsov, V., Birkinshaw, S., Olive, V., Lomax, J., Christie, D., & Arthur, S. (2022). Multiple benefits of blue-green infrastructure and the reduction of environmental risks: Case study of ecosystem services provided by a suds pond. In *Civil Engineering for Disaster Risk Reduction*. Springer, pp. 247–262.

Kundu, D., & Pandey, A. K. (2020). World urbanisation: Trends and patterns. In *Developing National Urban Policies*. Springer, pp. 13–49.

Lacasta, A. M., Peñaranda, A., & Cantalapiedra, I. R. (2018). Green streets for noise reduction. In *Nature Based Strategies for Urban and Building Sustainability*. Elsevier, pp. 181–190.

Lakes, T., & Kim, H.-O. (2012). The urban environmental indicator "Biotope Area Ratio" — An enhanced approach to assess and manage the urban ecosystem services using high resolution remote-sensing. *Ecological Indicators, 13*(1), 93–103.

Lamond, J., & Everett, G. (2019). Sustainable blue-green infrastructure: A social practice approach to understanding community preferences and stewardship. *Landscape and Urban Planning, 191*, 103639.

Landschaft Planen & Bauen, & Becker, Giseke, Mohren, Richard. (1990). The Biotope Area Factor as an ecological parameter. Principles for its determination and identification of the target. Senate Department for Urban Development in Berlin, Berlin.

Lausch, A., Blaschke, T., Haase, D., Herzog, F., Syrbe, R.-U., Tischendorf, L., & Walz, U. (2015). Understanding and quantifying landscape structure — A review on relevant process characteristics, data models and landscape metrics. *Ecological Modelling, 295*, 31–41.

Lee, D., Pietrzyk, P., Donkers, S., Liem, V., van Oostveen, J., Montazeri, S., ... Nerry, F. (2013). Modelling and observation of heat losses from buildings: The impact of geometric detail on 3D heat flux modelling. Paper presented at the Proceedings of the 33rd AERSel Symposium, Matera, Italy, 3–6 June 2013.

Lee, K. Y. (2012). *From Third World to First: The Singapore Story, 1965–2000*. Marshall Cavendish International Asia, Singapore.

Licina, D., & Yildirim, S. (2021). Occupant satisfaction with indoor environmental quality, sick building syndrome (SBS) symptoms and self-reported productivity before and after relocation into WELL-certified office buildings. *Building and Environment, 204*, 108183.

Lidmo, J., Bogason, Á., & Turunen, E. (2020). The legal framework and national policies for urban greenery and green values in urban areas. Nordregio Report.

Lohr, V., & Pearson-Mims, C. H. (2000). Physical discomfort may be reduced in the presence of interior plants. *HortTechnology*, *10*, 53–58. doi:10.21273/HORTTECH.10.1.53

Lohr, V., Pearson-Mims, C. H., & Goodwin, G. K. (1995). Interior plants may improve worker productivity and reduce stress in a windowless environment. *Journal of Environmental Horticulture, 14*.

Lyytimäki, J., Sipilä, M. J. U. F., & Greening, U. (2009). Hopping on one leg — The challenge of ecosystem disservices for urban green management. *8*(4), 309–315.

Lyytimäki, J., Petersen, L. K., Normander, B., & Bezák, P. (2008). Nature as a nuisance? Ecosystem services and disservices to urban lifestyle. *Environmental Sciences, 5*(3), 161–172.

MacIvor, J. S., & Lundholm, J. (2011). Insect species composition and diversity on intensive green roofs and adjacent level-ground habitats. *Urban Ecosystems, 14*(2), 225–241.

MacIvor, J. S., Margolis, L., Puncher, C. L., & Matthews, B. J. C. (2013). Decoupling factors affecting plant diversity and cover on extensive green roofs. *Journal of Environmental Management, 130*, 297–305.

Mahmoud, A. S., Asif, M., Hassanain, M. A., Babsail, M. O., & Sanni-Anibire, M. O. (2017). Energy and economic evaluation of green roofs for residential buildings in hot-humid climates. *Buildings, 7*(2), 30.

Manso, M., Teotónio, I., Silva, C. M., & Cruz, C. O. (2021). Green roof and green wall benefits and costs: A review of the quantitative evidence. *Renewable and Sustainable Energy Reviews, 135*, 110111.

Matteson, K. C., & Langellotto, G. A. (2017). Bumble bee abundance in New York City community gardens: Implications for urban agriculture. In *Urban Horticulture*. Apple Academic Press, pp. 217–234.

McCree, K. J. (1981). Photosynthetically active radiation. In *Physiological Plant Ecology I: Responses to the Physical Environment*, pp. 41–55.

McGarigal, K. (1995). *FRAGSTATS: Spatial Pattern Analysis Program for Quantifying Landscape Structure*. US Department of Agriculture, Forest Service, Pacific Northwest Research Station.

McIntyre, N. E., Knowles-Yánez, K., & Hope, D. (2008). Urban ecology as an interdisciplinary field: Differences in the use of "urban" between the social and natural sciences. In *Urban Ecology*. Springer, pp. 49–65.

McPhearson, P. T., Feller, M., Felson, A., Karty, R., Lu, J., Palmer, M. I., & Wenskus, T. (2017). Assessing the effects of the urban forest restoration effort of MillionTreesNYC on the structure and functioning of New York City ecosystems. *Urban Forests: Ecosystem Services and Management, 265*.

McPherson, E. G., & Simpson, J. R. (1999). *Carbon Dioxide Reduction Through Urban Forestry*. US Department of Agriculture, Forest Service, Pacific Southwest Research.

Medl, A., Stangl, R., & Florineth, F. (2017). Vertical greening systems — a review on recent technologies and research advancement. *Building and Environment, 125*, 227–239. doi:10.1016/j.buildenv.2017.08.054

Melbourne, City of. (2012). Urban forest strategy: Making a great city greener, 2012–2032. https://www.melbourne.vic.gov.au/SiteCollectionDocuments/urban-forest-strategy.pdf

Melbourne, City of. (2017). Green our city strategic action plan. https://www.melbourne.vic.gov.au/sitecollectiondocuments/green-our-city-action-plan-2018.pdf

Mickovski, S. B., Buss, K., McKenzie, B. M., & Sökmener, B. (2013). Laboratory study on the potential use of recycled inert construction waste material in the substrate mix for extensive green roofs. *Ecological Engineering, 61*, 706–714.

Mocca, E., Friesenecker, M., & Kazepov, Y. (2020). Greening Vienna. The multi-level interplay of urban environmental policy — making. *Sustainability, 12*(4), 1577.

Moir, L. (2001). What do we mean by corporate social responsibility? *Corporate Governance: The International Journal of Business in Society, 1*(2), 16–22.

Mok, W. K., Tan, Y. X., & Chen, W. N. (2020). Technology innovations for food security in Singapore: A case study of future food systems for an increasingly natural resource-scarce world. *Trends in Food Science & Technology, 102*, 155–168.

Moll, G., & Petit, J. (1994). The urban ecosystem: Putting nature back in the picture. *Urban Forests, 14*(5), 8–15.

Monteiro, M. V., Doick, K. J., Handley, P., & Peace, A. (2016). The impact of greenspace size on the extent of local nocturnal air temperature cooling in London. *Urban Forestry & Urban Greening, 16*, 160–169.

Monteith, J. (1972). Solar radiation and productivity in tropical ecosystems. *Journal of Applied Ecology, 9*(3), 747–766.

Morakinyo, T. E., Lau, K. K.-L., Ren, C., & Ng, E. (2018). Performance of Hong Kong's common trees species for outdoor temperature regulation, thermal comfort and energy saving. *Building and Environment, 137*, 157–170.

Moshiri, G. A. (2020). *Constructed Wetlands for Water Quality Improvement.* CRC Press.

Nagendra, H. (2001). Using remote sensing to assess biodiversity. *International Journal of Remote Sensing, 22*(12), 2377–2400.

Napoli, M., Massetti, L., Brandani, G., Petralli, M., & Orlandini, S. (2016). Modeling tree shade effect on urban ground surface temperature. *Journal of Environmental Quality, 45*(1), 146–156.

NEA. (2021). *Transforming Singapore into a City in Nature.* National Environment Agency, Singapore. https://www.nea.gov.sg/docs/default-source/default-document-library/annex-b---factsheet-on-city-in-nature.pdf

Nissim, W. G., & Labrecque, M. (2021). Reclamation of urban brownfields through phytoremediation: Implications for building sustainable and resilient towns. *Urban Forestry & Urban Greening, 65,* 127364.

Nowak, D. J., & Crane, D. E. (2002). Carbon storage and sequestration by urban trees in the USA. *Environmental Pollution, 116*(3), 381–389.

NParks. (2011). Launch of 27 container trees at Cashew Road — New approach in tree planting reduces number of trees affected by development. https://www.nparks.gov.sg/news/2011/1/launch-of-27-container-trees-at-cashew-road--new-approach-in-tree-planting-reduces-number-of-trees-affected-by-development

Oh, R. R., Richards, D. R., & Yee, A. T. (2018). Community-driven skyrise greenery in a dense tropical city provides biodiversity and ecosystem service benefits. *Landscape and Urban Planning, 169,* 115–123.

Olszewska-Guizzo, A., Escoffier, N., Chan, J., & Tan, P. Y. (2018). Window view and the brain: Effects of floor level and green cover on the alpha and beta rhythms in a passive exposure EEG experiment. *International Journal of Environmental Research and Public Health, 15*(11), 2358.

Olszewska-Guizzo, A., Sia, A., Fogel, A., & Ho, R. (2020). Can exposure to certain urban green spaces trigger frontal alpha asymmetry in the brain? Preliminary findings from a passive task EEG study. *International Journal of Environmental Research and Public Health, 17*(2), 394.

Olszewska-Guizzo, A., Mukoyama, A., Naganawa, S., Dan, I., Husain, S. F., Ho, C. S., & Ho, R. (2021). Hemodynamic response to three types of urban spaces before and after lockdown during the COVID-19 pandemic. *International Journal of Environmental Research and Public Health, 18*(11), 6118.

Olszewska-Guizzo, A., Fogel, A., Escoffier, N., Sia, A., Nakazawa, K., Kumagai, A., ... Ho, R. (2022). Therapeutic garden with contemplative features induces desirable changes in mood and brain activity in depressed adults. *Frontiers in Psychiatry, 13.*

Ong, B. L. (2003). Green plot ratio: An ecological measure for architecture and urban planning. *Landscape and Urban Planning, 63*(4), 197–211.

Ottelé, M., van Bohemen, H. D., & Fraaij, A. L. A. (2010). Quantifying the deposition of particulate matter on climber vegetation on living walls. *Ecological Engineering, 36*(2), 154–162. https://doi.org/10.1016/j.ecoleng.2009.02.007

Ouyang, W., Morakinyo, T. E., Ren, C., & Ng, E. (2020). The cooling efficiency of variable greenery coverage ratios in different urban densities: A study in a subtropical climate. *Building and Environment, 174,* 106772.

Ow, L. F., & Ghosh, S. (2017). Growth of street trees in urban ecosystems: Structural cells and structural soil. *Journal of Urban Ecology, 3*(1), jux017.

Page, J., Winston, R., & Hunt, W. H. III. (2014). Field monitoring of two Silva Cell™ installations in Wilmington, North Carolina. https://www. deeproot.com/silvapdfs/resources/articles/NCSU_deep_root_final_ report_1_27_2014.pdf

Panduro, T. E., & Veie, K. L. (2013). Classification and valuation of urban green spaces — A hedonic house price valuation. *Landscape and Urban Planning, 120*, 119–128.

Park, J.-C., Yang, H.-M., & Jang, B.-K. (2012). Policy for establishment of green infrastructure. *Journal of the Korean Institute of Landscape Architecture, 40*(5), 43–50.

Parks Forestry and Recreation City Planning, City of Toronto. (2019). Parkland strategy: Growing Toronto parkland. https://www.toronto.ca/wp-content/ uploads/2019/11/97fb-parkland-strategy-full-report-final.pdf

Pavlineri, N., Skoulikidis, N. T., & Tsihrintzis, V. A. (2017). Constructed floating wetlands: A review of research, design, operation and management aspects, and data meta-analysis. *Chemical Engineering Journal, 308*, 1120–1132.

Peri, G., Traverso, M., Finkbeiner, M., & Rizzo, G. (2012). The cost of green roofs disposal in a life cycle perspective: Covering the gap. *Energy, 48*(1), 406–414.

Perini, K., & Rosasco, P. (2013). Cost–benefit analysis for green façades and living wall systems. *Building and Environment, 70*, 110–121.

Perini, K., & Rosasco, P. (2016). Is greening the building envelope economically sustainable? An analysis to evaluate the advantages of economy of scope of vertical greening systems and green roofs. *Urban Forestry & Urban Greening, 20*, 328–337.

Perini, K., Ottelé, M., Haas, E., & Raiteri, R. (2011). Greening the building envelope, façade greening and living wall systems. *Open Journal of Ecology, 1*(01), 1.

Perini, K., Ottelé, M., Giulini, S., Magliocco, A., & Roccotiello, E. (2017). Quantification of fine dust deposition on different plant species in a vertical greening system. *Ecological Engineering, 100*, 268–276. https://doi. org/10.1016/j.ecoleng.2016.12.032

Peterson, M. E. (1980). Soil percolation tests. *Journal of Environmental Health*, 182–186.

Pettit, T., Irga, P., Abdo, P., & Torpy, F. (2017). Do the plants in functional green walls contribute to their ability to filter particulate matter? *Building and Environment, 125*, 299–307.

Pettorelli, N., Vik, J. O., Mysterud, A., Gaillard, J.-M., Tucker, C. J., & Stenseth, N. C. (2005). Using the satellite-derived NDVI to assess ecological responses to environmental change. *Trends in Ecology & Evolution, 20*(9), 503–510.

Piorr, A., Ravetz, J., & Tosics, I. (2011). Peri-urbanisation in Europe: Towards European policies to sustain urban-rural futures. Synthesis Report. https://www.openspace.eca.ed.ac.uk/wp-content/uploads/2015/12/Peri_ Urbanisation_in_Europe_printversion.pdf

Pisello, A. L., Piselli, C., & Cotana, F. (2015). Thermal-physics and energy performance of an innovative green roof system: The cool-green roof. *Solar Energy, 116*, 337–356.

Planning Department, Hong Kong. (2015). Recreation, open space and greening. In *Hong Kong Planning Standards And Guidelines*. The Government of the Hong Kong Special Administrative Region. https://www.pland.gov.hk/ pland_en/tech_doc/hkpsg/full/pdf/ch4.pdf

Planning Department, Hong Kong. (2021). About Hong Kong 2030+. The Government of the Hong Kong Special Administrative Region. https://www. pland.gov.hk/pland_en/p_study/comp_s/hk2030plus/about_a.htm

PUB. (2014). *Active Beautiful Clean Waters: Design Guidelines*. PUB Singapore.

Punmia, B., & Jain, A. K. (2005). *Soil Mechanics and Foundations*. Firewall Media.

Rahoui, H. (2021). Greenest City 2020, Vancouver. In *Urban Planning for Transitions*, pp. 47–67.

Røe, P. G. (2016). *Green Oslo: Visions, Planning and Discourse*. Routledge.

Ruiz-Aviles, V., Brazel, A., Davis, J. M., & Pijawka, D. (2020). Mitigation of Urban Heat Island effects through "Green Infrastructure": Integrated design of constructed wetlands and neighborhood development. *Urban Science, 4*(4), 78.

Runkle, E., & Bugbee, B. (2013). Problems with foot-candles, lux and lumens. Michigan State University. https://www.canr.msu.edu/uploads/resources/ pdfs/footcandles-lux-lumens2.pdf

Saadatian, O., Sopian, K., Salleh, E., Lim, C., Riffat, S., Saadatian, E., ... Sulaiman, M. (2013). A review of energy aspects of green roofs. *Renewable and Sustainable Energy Reviews, 23*, 155–168.

Sadineni, S. B., Madala, S., & Boehm, R. F. (2011). Passive building energy savings: A review of building envelope components. *Renewable and Sustainable Energy Reviews, 15*(8), 3617–3631.

Safikhani, T., Abdullah, A. M., Ossen, D. R., & Baharvand, M. (2014). A review of energy characteristic of vertical greenery systems. *Renewable and Sustainable Energy Reviews, 40*, 450–462.

Salata, F., Golasi, I., de Lieto Vollaro, R., & de Lieto Vollaro, A. (2016). Urban microclimate and outdoor thermal comfort. A proper procedure to fit ENVI-met simulation outputs to experimental data. *Sustainable Cities and Society, 26*, 318–343.

Samsudin, R., Yok, T. P., & Chua, V. (2022). Social capital formation in high density urban environments: Perceived attributes of neighborhood green space shape social capital more directly than physical ones. *227*, 104527.

Sanglerat, G. (2012). *The Penetrometer and Soil Exploration*. Elsevier.

Santamouris, M. (2014). Cooling the cities — a review of reflective and green roof mitigation technologies to fight heat island and improve comfort in urban environments. *Solar Energy, 103*, 682–703.

Santamouris, M. (2015). Analyzing the heat island magnitude and characteristics in one hundred Asian and Australian cities and regions. *Science of the Total Environment, 512*, 582–598.

Schröpfer, T., & Menz, S. (2019). Frameworks and guidelines promoting high-rise greenery. In *Dense and Green Building Typologies*. Springer, pp. 13–19.

Seattle Parks Foundation. (2014). South Park Green Space Vision Plan. https://www.seattle.gov/documents/Departments/Environment/EnvironmentalEquity/South-Park-Green-Space-Vision-Plan_6.17.14_Final-with-Appendix.pdf

Şekercioğlu, Ç. H., Daily, G. C., & Ehrlich, P. R. (2004). Ecosystem consequences of bird declines. *Proceedings of the National Academy of Sciences, 101*(52), 18042–18047.

Setiowati, R., Hasibuan, H., & Koestoer, R. (2018). Green open space masterplan at Jakarta Capital City, Indonesia for climate change mitigation. Paper presented at the IOP Conference Series: Earth and Environmental Science.

Sev, A. (2009). How can the construction industry contribute to sustainable development? A conceptual framework. *Sustainable Development, 17*(3), 161–173.

Seyam, S. (2019). The impact of greenery systems on building energy: Systematic review. *Journal of Building Engineering, 26*, 100887.

Shafique, M., Kim, R., & Rafiq, M. (2018). Green roof benefits, opportunities and challenges — A review. *Renewable and Sustainable Energy Reviews, 90*, 757–773.

Sharifi, A. (2016). From Garden City to eco-urbanism: The quest for sustainable neighborhood development. *Sustainable Cities and Society, 20*, 1–16.

Shentova, R., De Vries, S., & Verboom, J. (2022). Well-being in the time of Corona: Associations of nearby greenery with mental well-being during COVID-19 in the Netherlands. *Sustainability, 14*(16), 10256.

Singapore Green Plan. (2021). Singapore Green Plan 2030 charts ambitious targets for 10 years to catalyse national sustainability movement. https://www.greenplan.gov.sg/news/press-releases/2021-02-10-press-release-on-green-plan

SITES. (2022). Sites rating system. https://sustainablesites.org/certification-guide

Soane, B. D., & van Ouwerkerk, C. (2013). *Soil Compaction in Crop Production*. Elsevier.

Song, X. P., Tan, H. T., & Tan, P. Y. (2018). Assessment of light adequacy for vertical farming in a tropical city. *Urban Forestry & Urban Greening, 29,* 49–57.

Soreanu, G., Dixon, M., & Darlington, A. (2013). Botanical biofiltration of indoor gaseous pollutants — a mini-review. *Chemical Engineering Journal, 229,* 585–594.

Soriano, A., & Paruelo, J. M. (1992). Biozones: Vegetation units defined by functional characters identifiable with the aid of satellite sensor images. *Global Ecology and Biogeography Letters,* pp. 82–89.

Spahr, K. M., Bell, C. D., McCray, J. E., & Hogue, T. S. (2020). Greening up stormwater infrastructure: Measuring vegetation to establish context and promote cobenefits in a diverse set of US cities. *Urban Forestry & Urban Greening, 48,* 126548.

Sproul, J., Wan, M. P., Mandel, B. H., & Rosenfeld, A. H. (2014). Economic comparison of white, green, and black flat roofs in the United States. *Energy and Buildings, 71,* 20–27.

Stansfeld, S., Haines, M., & Brown, B. (2000). Noise and health in the urban environment. *Reviews on Environmental Health, 15*(1–2), 43–82.

Steele, J. (1997). *Sustainable Architecture: Principles, Paradigms, and Case Studies*. McGraw-Hill.

Sternberg, T., Viles, H., Cathersides, A., & Edwards, M. (2010). Dust particulate absorption by ivy (Hedera helix L) on historic walls in urban environments. *Science of the Total Environment, 409*(1), 162–168. doi:https://doi.org/10.1016/j.scitotenv.2010.09.022

Sydney, City of. (2021). Greening Sydney strategy. https://www.cityofsydney.nsw.gov.au/-/media/corporate/files/publications/strategies-action-plans/greening-sydney-strategy/greening-sydney-strategy.pdf?download=true

Szibbo, N. (2016). Lessons for LEED® for neighborhood development, social equity, and affordable housing. *Journal of the American Planning Association, 82*(1), 37–49.

Tan, B. A., Gaw, L. Y.-F., Masoudi, M., & Richards, D. R. (2021). Nature-based solutions for urban sustainability: An ecosystem services assessment of plans for Singapore's first "forest town". *Frontiers in Environmental Science, 9*, 610155.

Tan, C. (2022). LTA, NParks plant a cool idea at bus stops islandwide. *The Straits Times*. https://www.straitstimes.com/singapore/transport/lta-nparks-plant-a-cool-idea-at-bus-stops-islandwide

Tan, C. L., Wong, N. H., Tan, P. Y., Ismail, M., & Wee, L. Y. (2017). Growth light provision for indoor greenery: A case study. *Energy and Buildings, 144*, 207–217.

Tan, C. L., Wong, N. H., Tan, P. Y., Jusuf, S. K., & Chiam, Z. Q. (2015). Impact of plant evapotranspiration rate and shrub albedo on temperature reduction in the tropical outdoor environment. *Building and Environment, 94*, 206–217.

Tan, C. L., Tan, P. Y., Wong, N. H., Takasuna, H., Kudo, T., Takemasa, Y., … Chua, H. X. V. (2017). Impact of soil and water retention characteristics on green roof thermal performance. *Energy and Buildings, 152*, 830–842.

Tan, P. Y. (2017). Greening Singapore: Past achievements, emerging challenges. In *50 Years of Urban Planning in Singapore*. World Scientific Publishing, Singapore, pp. 177–195.

Tan, P. Y., & Ismail, M. R. B. (2014). Building shade affects light environment and urban greenery in high-density residential estates in Singapore. *Urban Forestry & Urban Greening, 13*(4), 771–784.

Tan, P. Y., & Ismail, M. R. B. (2016). Photosynthetically active radiation and comparison of methods for its estimation in equatorial Singapore. *Theoretical and Applied Climatology, 123*(3–4), 873–883.

Tan, P. Y., & Sia, A. (2010). Leaf area index of tropical plants: A guidebook on its use in the calculation of green plot ratio. Centre of Urban Greenery and Ecology, Singapore.

Tan, P. Y., Wang, J., & Sia, A. (2013). Perspectives on five decades of the urban greening of Singapore. *Cities, 32*, 24–32.

Tan, P. Y., Liao, K.-h., Hwang, Y. H., & Chua, V. (2018). *Nature, Place & People: Forging Connections Through Neighbourhood Landscape Design*. World Scientific Publishing, Singapore.

Tariku, F., & Hagos, S. (2022). Performance of green roof installed on highly insulated roof deck and the plants' effect: An experimental study. *Building and Environment*, 109337.

Teshnehdel, S., Akbari, H., Di Giuseppe, E., & Brown, R. D. (2020). Effect of tree cover and tree species on microclimate and pedestrian comfort in a residential district in Iran. *Building and Environment, 178*, 106899.

Thaiutsa, B., Puangchit, L., Kjelgren, R., & Arunpraparut, W. (2008). Urban green space, street tree and heritage large tree assessment in Bangkok, Thailand. *Urban Forestry & Urban Greening, 7*(3), 219–229.

The Government of the Hong Kong Special Administrative Region. (nd). GMP for urban areas. https://www.cedd.gov.hk/eng/topics-in-focus/greening/urban/index.html

The People's Government of Beijing Municipality. (2020). Report on the Work of the Government 2020 (part III). http://english.beijing.gov.cn/government/reports/202005/t20200510_1893589.html

Thien, S. J. (1979). A flow diagram for teaching texture by feel analysis. *Journal of Agronomic Education, 8*(1), 54–55.

Thierfelder, H., & Kabisch, N. (2016). Viewpoint Berlin: Strategic urban development in Berlin — Challenges for future urban green space development. *Environmental Science & Policy, 62*, 120–122.

Torres, A. P., & Lopez, R. G. (2010). Measuring daily light integral in a greenhouse. Purdue Extension, West Lafayette, p. 7.

Townshend, D., & Duggie, A. (2007). Study on green roof application in Hong Kong. Architectural Services Department.

Trojanek, R., Gluszak, M., & Tanas, J. J. (2018). The effect of urban green spaces on house prices in Warsaw. *International Journal of Strategic Property Management, 22*(5), 358–371.

Trowbridge, P. J., & Bassuk, N. L. (2004). *Trees in the Urban Landscape: Site Assessment, Design, and Installation*. John Wiley & Sons.

Troy, A., & Wilson, M. A. (2006). Mapping ecosystem services: Practical challenges and opportunities in linking GIS and value transfer. *Ecological Economics, 60*(2), 435–449.

URA. (2017). Updates to the landscaping for urban spaces and high-rises (LUSH) programme: LUSH 3.0. Urban Redevelopment Authority, Singapore.

USDA, N. S. (1993). Soil texture calculator. US Department of Agriculture. https://www.nrcs.usda. gov/wps/portal/nrcs/detail/soils/survey.

USGBC. (2022). LEED v4.1 Building design and construction. US Green Building Council. https://build.usgbc.org/bdc41

Uuemaa, E., Antrop, M., Roosaare, J., Marja, R., & Mander, Ü. (2009). Landscape metrics and indices: An overview of their use in landscape research. *Living Reviews in Landscape Research, 3*(1), 1–28.

Van Renterghem, T., & Botteldooren, D. (2011). In-situ measurements of sound propagating over extensive green roofs. *Building and Environment, 46*(3), 729–738.

Van Renterghem, T., Botteldooren, D., & Verheyen, K. (2012). Road traffic noise shielding by vegetation belts of limited depth. *Journal of Sound and Vibration, 331*(10), 2404–2425. doi:https://doi.org/10.1016/j.jsv.2012.01.006

Vancouver, City of. (2020). Green Vancouver — Climate Emergency Action Plan. https://vancouver.ca/files/cov/climate-emergency-action-plan-summary.pdf

VanDerZanden, A. M., & Cook, T. W. (2010). *Sustainable Landscape Management: Design, Construction, and Maintenance.* John Wiley & Sons.

Velasco, E., & Roth, M. (2010). Cities as net sources of CO_2: Review of atmospheric CO_2 exchange in urban environments measured by eddy covariance technique. *Geography Compass, 4*(9), 1238–1259.

Velasco, E., Roth, M., Norford, L., & Molina, L. T. (2016). Does urban vegetation enhance carbon sequestration? *Landscape and Urban Planning, 148,* 99–107.

Vieira, J., Matos, P., Mexia, T., Silva, P., Lopes, N., Freitas, C., … Pinho, P. (2018). Green spaces are not all the same for the provision of air purification and climate regulation services: The case of urban parks. *Environmental Research, 160,* 306–313.

Virtudes, A. (2016). Benefits of greenery in contemporary city. Paper presented at the IOP Conference Series: Earth and Environmental Science.

Vymazal, J. (2011). Constructed wetlands for wastewater treatment: Five decades of experience. *Environmental Science & Technology, 45*(1), 61–69.

Weier, J., & Herring, D. (2000). Measuring vegetation (NDVI & EVI). Earth Observatory, National Aeronautics and Space Administration.

WELL. (2022). The WELL Building Standard. WELL v2 Q4. https://v2.wellcertified.com/en/wellv2/overview

Well, F., & Ludwig, F. (2020). Blue–green architecture: A case study analysis considering the synergetic effects of water and vegetation. *Frontiers of Architectural Research, 9*(1), 191–202.

Wellington City Council. (2015). Our natural capital: Wellington's Biodiversity Strategy and Action Plan 2015. Wellington City Council.

Werner, P. (2011). The ecology of urban areas and their functions for species diversity. *Landscape and Ecological Engineering, 7*(2), 231–240.

Wilson, L. (2009). Observations and suggestions regarding the proposed Hanoi capital construction Master Plan to 2030 and vision to 2050. Paper presented at the International Symposium for the Hanoi Capital Construction Master Plan to 2030 and Vision to 2050.

Wolsink, M. (2016). Environmental education excursions and proximity to urban green space–densification in a "compact city". *Environmental Education Research, 22*(7), 1049–1071.

Wong, N. H., Tan, P. Y., & Chen, Y. (2007). Study of thermal performance of extensive rooftop greenery systems in the tropical climate. *Building and Environment, 42*(1), 25–54.

Wong, N. H., Tan, A. Y. K., Tan, P. Y., & Wong, N. C. (2009). Energy simulation of vertical greenery systems. *Energy and Buildings, 41*(12), 1401–1408.

Wong, N. H., Tan, C. L., Kolokotsa, D. D., & Takebayashi, H. (2021). Greenery as a mitigation and adaptation strategy to urban heat. *Nature Reviews Earth & Environment, 2*(3), 166–181.

Wong, N. H., Jusuf, S. K., Syafii, N. I., Chen, Y., Hajadi, N., Sathyanarayanan, H., & Manickavasagam, Y. V. (2011). Evaluation of the impact of the surrounding urban morphology on building energy consumption. *Solar Energy, 85*(1), 57–71.

Wong, N. H., Tan, A. Y. K., Tan, P. Y., Chiang, K., & Wong, N. C. (2010). Acoustics evaluation of vertical greenery systems for building walls. *Building and Environment, 45*(2), 411–420.

Wong, N. H., Tay, S. F., Wong, R., Ong, C. L., & Sia, A. (2003). Life cycle cost analysis of rooftop gardens in Singapore. *Building and Environment, 38*(3), 499–509.

Wong, N. H., Jusuf, S. K., La Win, A. A., Thu, H. K., Negara, T. S., & Xuchao, W. (2007). Environmental study of the impact of greenery in an institutional campus in the tropics. *Building and Environment, 42*(8), 2949–2970.

Wooster, E., Fleck, R., Torpy, F., Ramp, D., & Irga, P. (2022). Urban green roofs promote metropolitan biodiversity: A comparative case study. *Building and Environment, 207*, 108458.

Wu, L., & Kim, S. K. (2021). Does socioeconomic development lead to more equal distribution of green space? Evidence from Chinese cities. *Science of the Total Environment, 757*, 143780. https://doi.org/10.1016/j.scitotenv.2020.143780

Xi, C., Ding, J., Wang, J., Feng, Z., & Cao, S.-J. (2022). Nature-based solution of greenery configuration design by comprehensive benefit evaluation of microclimate environment and carbon sequestration. *Energy and Buildings, 270*, 112264.

Xie, L., Flynn, A., Tan-Mullins, M., & Cheshmehzangi, A. (2019). The making and remaking of ecological space in China: The political ecology of Chongming Eco-Island. *Political Geography, 69*, 89–102.

Yang, J., Yu, Q., & Gong, P. (2008). Quantifying air pollution removal by green roofs in Chicago. *Atmospheric Environment, 42*(31), 7266–7273.

Yang, J., Zhao, L., Mcbride, J., & Gong, P. (2009). Can you see green? Assessing the visibility of urban forests in cities. *Landscape and Urban Planning, 91*(2), 97–104.

Yang, L., Zhang, L., Ma, J., Xie, J., & Liu, L. (2011). Interactive visualization of multi-resolution urban building models considering spatial cognition. *International Journal of Geographical Information Science, 25*(1), 5–24.

Yeom, S., Kim, H., & Hong, T. (2021). Psychological and physiological effects of a green wall on occupants: A cross-over study in virtual reality. *Building and Environment, 204*, 108134.

Yudelson, J., & Meyer, U. (2013). *The World's Greenest Buildings: Promise Versus Performance in Sustainable Design.* Routledge.

Zhang, F., Wu, F., & Lin, Y. (2022). The socio-ecological fix by multi-scalar states: The development of 'Greenways of Paradise' in Chengdu. *Political Geography, 98*, 102736. https://doi.org/10.1016/j.polgeo.2022.102736

Zhang, Y., Zhang, T., Zeng, Y., Yu, C., & Zheng, S. (2021). The rising and heterogeneous demand for urban green space by Chinese urban residents: Evidence from Beijing. *Journal of Cleaner Production, 313*, 127781.

Zhang, Z., Szota, C., Fletcher, T. D., Williams, N. S., Werdin, J., & Farrell, C. (2018). Influence of plant composition and water use strategies on green roof stormwater retention. *Science of the Total Environment, 625*, 775–781.

Zhao, M., Tabares-Velasco, P. C., Srebric, J., Komarneni, S., & Berghage, R. (2014). Effects of plant and substrate selection on thermal performance of green roofs during the summer. *Building and Environment, 78*, 199–211.

Zhao, Y., Chen, Y., & Li, K. (2022). A simulation study on the effects of tree height variations on the façade temperature of enclosed courtyard in North China. *Building and Environment, 207*, 108566.

Zheng, S., Guldmann, J.-M., Liu, Z., Zhao, L., Wang, J., Pan, X., & Zhao, D. (2020). Predicting the influence of subtropical trees on urban wind through wind tunnel tests and numerical simulations. *Sustainable Cities and Society, 57*, 102116.

Zingoni de Baro, M. E. (2022). Curitiba case study. In *Regenerating Cities: Reviving Places and Planet.* Springer, pp. 117–162.

Index

www.ingramcontent.com/pod-product-compliance
Lightning Source LLC
Chambersburg PA
CBHW050602190326
41458CB00007B/2140